Fast Start Differential Calculus

Synthesis Lectures on Mathematics and Statistics

Editor
Steven G. Kranz, *Washington University, St. Louis*

Fast Start Differential Calculus
Daniel Ashlock

ISBN: 978-3-031-01292-1   paperback
ISBN: 978-3-031-02420-7   ebook
ISBN: 978-3-031-00266-3   hardcover

DOI 10.1007/978-3-031-02420-7

A Publication in the Springer series
SYNTHESIS LECTURES ON MATHEMATICS AND STATISTICS

Lecture #28
Series Editor: Steven G. Kranz, *Washington University, St. Louis*
Series ISSN
Print 1938-1743   Electronic 1938-1751

Fast Start Differential Calculus

Daniel Ashlock
University of Guelph

SYNTHESIS LECTURES ON MATHEMATICS AND STATISTICS #28

ABSTRACT

This book reviews the algebraic prerequisites of calculus, including solving equations, lines, quadratics, functions, logarithms, and trig functions. It introduces the derivative using the limit-based definition and covers the standard function library and the product, quotient, and chain rules. It explores the applications of the derivative to curve sketching and optimization and concludes with the formal definition of the limit, the squeeze theorem, and the mean value theorem.

KEYWORDS

differential calculus, review of algebra, curve sketching, optimization, limits

Contents

Preface

This text covers single-variable differential calculus and reviews necessary algebra skills. It was developed for a course that arose from a perennial complaint by the physics department at the University of Guelph that the introductory calculus courses covered topics roughly a year after they were needed. In an attempt to address this concern, a multi-disciplinary team created a two-semester integrated calculus and physics course. This book covers the differential calculus topics from that course. The philosophy of the course was that the calculus be delivered before it is needed, often just in time, and that the physics serves as a substantial collection of motivating examples that anchor the student's understanding of the mathematics.

The course ran three times before this text was started, and it was used in draft form for the fourth offering of the course and then for two additional years. There is a good deal of classroom experience and testing behind this text. There is also enough information to confirm our hypothesis that the course would help students. The combined drop and flunk rate for this course is consistently under 3%, where 20% is more typical for first-year university calculus. Co-instruction of calculus and physics works. It is important to note that we did not achieve these results by watering down the math. The topics covered, in two semesters, are about half as many as are covered by a standard first-year calculus course. That's the big surprise: covering more topics faster increased the average grade and reduced the failure rate. Using physics as a knowledge anchor worked even better than we had hoped.

This text and its two companion volumes, *Fast Start Integral Calculus* and *Fast Start Advanced Calculus*, multivariate calculus make a number of innovations that have caused mathematical colleagues to raise objections. In mathematics it is traditional, even dogmatic, that math be taught in an order in which nothing is presented until the concepts on which it rests are already in hand. This is correct, useful dogma for mathematics students. It also leads to teaching difficult proofs to students who are still hungover from beginning-of-semester parties. This text neither emphasizes nor neglects theory, but it does move theory away from the beginning of the course in acknowledgment of the fact that this material is philosophically difficult and intellectually challenging. The course presents a broad integrated picture as soon as possible. Cleverness and computational efficiency are emphasized following the philosophy that "mathematics is the art of avoiding calculation."

The material on the mathematical foundations of limits was presented near the end of the first semester in the course for which these texts were developed. It appeared about mid-way through the original single-volume edition of the book. The partitioning of the original text into three parts left us with the question of where to put the theoretical material on limits and continuity. We settled on the end of this text, on the differential calculus. In the past, formal limits and continuity have been taught at the end of what is now the second text, on the integral calculus. Instructors can teach the material where they want—other than an appropriate level of mathematical maturity, there is no prerequisite for this material.

It is important to state what was sacrificed to make this course and this text work the way they do. This is not a good text for math majors, unless they get the theoretical parts of calculus later in a real analysis course. The text is relatively informal, almost entirely example driven, and application motivated. The author is a math professor with a CalTech Ph.D. and three decades of experience teaching math at all levels from 7th grade (as a volunteer) to graduate education including having supervised a dozen successful doctoral students. The author's calculus credits include calculus for math and engineering, calculus for biology, calculus for business, and multivariate and vector calculus.

Daniel Ashlock
August 2019

Acknowledgments

This text was written for a course developed by a team including my co-developers Joanne M. O'Meara of the Department of Physics and Lori Jones and Dan Thomas of the Department of Chemistry, University of Guelph. Andrew McEachern, Cameron McGuinness, Jeremy Gilbert, and Amanda Saunders have served as head TAs and instructors for the course over the last six years and had a substantial impact on the development of both the course and this text. Martin Williams, of the Department of Physics at Guelph, has been an able partner on the physics side, delivering the course and helping get the integration of the calculus and physics correct. I also owe six years of students thanks for serving as the test bed for the material. Many thanks to all these people for making it possible to decide what went into the text and what didn't. I also owe a great debt to Wendy Ashlock and Cameron McGuinness at Ashlock and McGuinness Consulting for removing a large number of errors and making numerous suggestions to enhance the clarity of the text.

Daniel Ashlock
August 2019

C H A P T E R   1

Review of Algebra

This book is a text on calculus, structured to prepare students for applying calculus to the physical sciences. The first chapter has no calculus in it at all; it is here because many students manage to get to the university or college level without adequate skill in algebra, trigonometry, or geometry. We assume familiarity with the concept of **variables** like x and y that denote numbers whose value is not known.

1.1   SOLVING EQUATIONS

An **equation** is an expression with an equals sign in it. For example:

$$x = 3$$

is a very simple equation. It tells us that the value of the variable x is the number 3. There are a number of rules we can use to manipulate equations. What these rules do is change an equation into another equation that has the same meaning but a different form. The things we can do to an equation without changing its meaning include the following.

- Add or subtract the same term from both sides. If that term is one that is already present in the equation, we may call this **moving the term to the other side**. When this happens, the term changes sign, from positive to negative or negative to positive. For example:

$$x - 4 = 5 \qquad \text{This is the original equation}$$
$$x = 5 + 4 \qquad \text{Add 4 to both sides (move 4 to the other side)}$$
$$x = 9 \qquad \text{Finish the arithmetic}$$

- Multiply or divide both sides by the same expression.

For example:

$$3x - 4 = 5 \qquad \text{This is the original equation}$$
$$3x = 5 + 4 \qquad \text{Add 4 to both sides (move 4 to the other side)}$$
$$3x = 9 \qquad \text{Finish the addition}$$
$$\frac{3x}{3} = \frac{9}{3} \qquad \text{Divide both sides by 3}$$
$$x = 3 \qquad \text{Finish the arithmetic}$$

- Apply the same function or operation to both sides.
  For example:

$$\sqrt{x - 2} = 5 \qquad \text{This is the original equation}$$
$$\left(\sqrt{x - 2}\right)^2 = 5^2 \qquad \text{Square both sides}$$
$$x - 2 = 25 \qquad \text{Do the arithmetic}$$
$$x = 27 \qquad \text{Move 2 to the other side}$$

Knowledge Box 1.1

The rules for solving equations include:

1. *Adding or subtracting the same thing from both sides.*

2. *Moving a term to the other side; its sign changes.*

3. *Multiplying or dividing both sides by the same thing.*

4. *Performing the same operation to both sides, e.g., squaring or taking the square root.*

These rules are illustrated by the examples in this section.

Solving an equation can get very hard at times and much of what we will do in this chapter is to give you tools for solving equations efficiently.

Example 1.1   Solve the equation $3x - 7 = x + 3$ for x.

Solution:

$$3x - 7 = x + 3 \qquad \text{This is the original equation}$$
$$2x = 10 \qquad \text{Move 7 and } x \text{ to the other side}$$
$$x = 5 \qquad \text{Divide both sides by 2}$$

$\Diamond$

All the examples so far have been solving for the variable x. We can solve for any symbol.

Example 1.2   Solve $PV = nRT$ for P.

Solution:

$$PV = nRT \qquad \text{This is the original equation}$$
$$P = \frac{nRT}{V} \qquad \text{Divide both sides by } V$$

$\Diamond$

When we solve an equation for a variable, we put it in **functional form**. We will learn more about functions in Section 1.4. When an equation is in functional form, the variable we solved for represents a value we are trying to find. In Example 1.2, which is the Ideal Gas Law, this would be the pressure of a gas. We call this the **dependent variable**. The variable (or variables) on the other side of the equation represent values that are changing. We call these the **independent variables**. In Example 1.2, these are n and T, the amount and temperature of the gas. (R is the universal gas constant—not a variable.)

Let's try a slightly harder example.

Example 1.3   Solve $y = \dfrac{2x - 1}{x + 2}$ for x.

Solution:

$$y = \frac{2x - 1}{x + 2}$$ This is the original equation

$$y(x + 2) = 2x - 1$$ Multiply both sides by $(x + 2)$

$$xy + 2y = 2x - 1$$ Distribute y

$$xy + 2y + 1 = 2x$$ Move 1 to the other side

$$2y + 1 = 2x - xy$$ Move xy to the other side

$$2y + 1 = x(2 - y)$$ Factor out x on the right

$$\frac{2y + 1}{2 - y} = x$$ Divide both sides by $(2 - y)$

$$x = \frac{2y + 1}{2 - y}$$ Neaten up

◇

Example 1.3 used a strategy. First, clear the denominator; second, get everything with x in it on one side and everything else on the other side; third, factor to get a single x; and fourth, divide by whatever is multiplied by x in order to isolate x, obtaining the solution.

Example 1.4   Solve $x^2 + y^2 = 25$ for y.

Solution:

$$x^2 + y^2 = 25$$ This is the original equation

$$y^2 = 25 - x^2$$ Move x^2 to the other side

$$y = \pm\sqrt{25 - x^2}$$ Take the square root of both sides

◇

The original equation in Example 1.4 graphs as a circle of radius 5 centered at the origin. When we take the square root of both sides, the fact that squaring a number makes it positive means that the plus and minus square roots are both correct. By convention, if we need a single expression, we take the positive square root.

PROBLEMS

Problem 1.5   Solve the following equations for x.

1. $7x + 4 = -17$

2. $8x - 6 = 66$

3. $7x + 5 = 68$

4. $5x + 3 = 43$

5. $5x + 4 = 19$

6. $2x - 5 = -9$

Problem 1.6   Solve the following equations for x.

1. $\sqrt{6x + 3} = 15$

2. $\sqrt{7x + 8} = 13$

3. $\sqrt{32x + 4} = 6$

4. $\sqrt{x - 1} = 10$

5. $\sqrt{39x + 3} = 9$

6. $\sqrt{2x + 3} = 19$

Problem 1.7   Solve the following equations for y.

1. $x^2 + 14y^2 = 4$

2. $2x^2 + 10y^2 = 3$

3. $2x^2 + 10y^2 = 20$

4. $8x^2 + 17y^2 = 16$

5. $8x^2 + 18y^2 = 9$

6. $15x^2 + 12y^2 = 8$

Problem 1.8   Solve the following equations for x.

1. $y = \dfrac{7x + 6}{4x + 3}$

3. $y = \dfrac{4x + 2}{3x + 3}$

5. $y = \dfrac{4x - 3}{4x + 4}$

2. $y = \dfrac{9x - 6}{x + 1}$

4. $y = \dfrac{4x - 3}{9x + 9}$

6. $y = \dfrac{2x - 1}{5x + 4}$

Problem 1.9   Solve for i: $Ni = Kq$

Problem 1.14   Solve for i: $\dfrac{1}{ni} = \dfrac{XJ}{e}$

Problem 1.10   Solve for a: $mua = n$

Problem 1.11   Solve for K: $\dfrac{1}{o} = \dfrac{KO}{r}$

Problem 1.15   Solve for n: $YV = \dfrac{1}{gn}$

Problem 1.12   Solve for U: $\dfrac{IL}{Fc} = U$

Problem 1.13   Solve for z: $\dfrac{1}{D} = \dfrac{nM}{z}$

Problem 1.16   Solve for U: $\dfrac{F}{D} = \dfrac{1}{Un}$

1.2   LINES

We will start with an example of a very simple line: $y = x + 1$ (Figure 1.1). For any value of x we add one to x and get y. We can draw the line by picking any two points on it, lining up a straight edge on the two points, and drawing along the straight edge.

The line $y = x + 1$ is made of all the points that look like $(x, x + 1)$. Five points like that have been plotted in Figure 1.1: (-2,-1), (-1,0), (0,1), (1,2), and (2,3).

Definition 1.1 *A line can always be written in the form* $y = mx + b$ *where m is the* **slope** *of the line, and b is the* **y-intercept** *of the line. Both m and b are numbers.*

There are different ways to interpret the parameters m and b.

- The number m is the amount the y-value increases for each increase of 1 in the x-value.

- If we pick two points on the line, then the **run** between those points is the change in the x-value, and the **rise** is the change in the y-value. The **slope**, m, is the rise divided by the run.

- The **y-intercept** b is where the line hits the y-axis. This means the point $(0, b)$ is on the line.

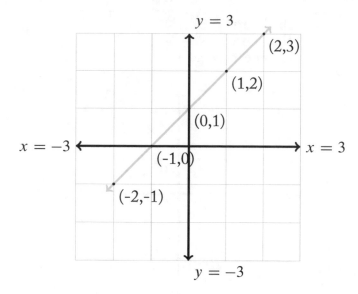

Figure 1.1: The line $y = x + 1$.

Let's look at what happens when we have lines with different slopes (Figure 1.2) or different intercepts (Figure 1.3).

1.2.1 COMPUTING SLOPES, FINDING LINES FROM POINTS

If someone gives us the slope, m, and the intercept, b, for a line and asks us the equation of the line, then it is easy, $y = mx + b$. This form is called the **slope-intercept** form of a line. It is the standard form for reporting a line—if you're asked to "find" a line, then you find it in slope-intercept form unless the directions specifically request a different form.

Knowledge Box 1.2

A line with slope m and y-intercept b has the equation:

$$y = mx + b$$

This is called **slope-intercept** *form. It is the usual form for reporting a line.*

A common state of affairs is to be given the slope of a line and a point that is on the line. This is one of the standard forms of a line and is easy to transform into slope-intercept form. It has

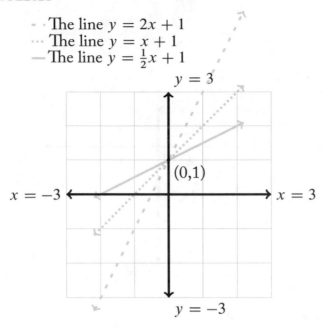

The line $y = 2x + 1$
The line $y = x + 1$
The line $y = \frac{1}{2}x + 1$

Figure 1.2: Changing slopes; same intercept.

the very sensible name of **point-slope form**.

Knowledge Box 1.3

A line with slope m containing the point (x_0, y_0) has the equation:

$$(y - y_0) = m(x - x_0)$$

This is called the **point-slope** *form of a line.*

Example 1.17   Find the line with slope $m = 2$ that passes through the point $(4, 4)$.

Solution:

$$y - y_0 = m(x - x_0) \qquad \text{Write the point-slope form}$$
$$y - 4 = 2(x - 4) \qquad \text{Put in the numbers}$$
$$y - 4 = 2x - 8 \qquad \text{Distribute the slope}$$
$$y = 2x - 4 \qquad \text{Add 4 to both sides}$$

So we see that the line (in slope-intercept form) is $y = 2x - 4$.

One of the things that you learn in elementary geometry is that **two points define a line**. This statement, while true, is a little short on details. Now that we have the point-slope form of a line, in order to find the equation of a line from two points, we need a way to find the slope from the two points. To do this, we need the fact that slope is rise over run. Given two points, we compute the rise and the run for those two points and then divide.

Knowledge Box 1.4

Suppose that we have two points (x_0, y_0) and (x_1, y_1) and want to know the slope of the line that passes through both of them. If $x_0 \neq x_1$ then the slope of the line is

$$m = \frac{y_1 - y_0}{x_1 - x_0}$$

This is the **slope of a line through two points**.

Example 1.18   Find the slope of the line through the points $(1, 1)$ and $(3, 4)$.

Solution:

$$m = \frac{y_1 - y_0}{x_1 - x_0} = \frac{4 - 1}{3 - 1} = \frac{3}{2}$$

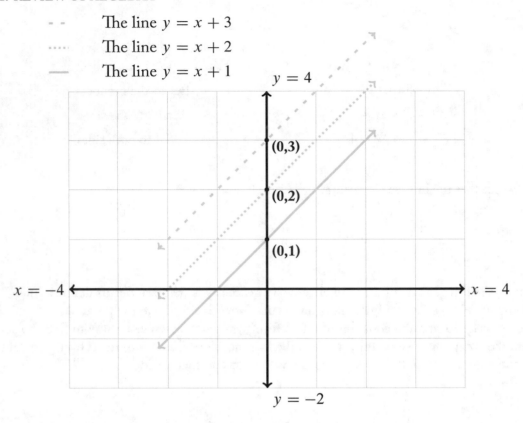

- - The line $y = x + 3$

.... The line $y = x + 2$

—— The line $y = x + 1$

Figure 1.3: Changing intercepts; same slopes.

◊

Now that we have a method for getting the slope of a line through two points, we can perform the slightly more complex task of finding the line through two points.

Example 1.19   Find the line through the points $(-2, 3)$ and $(2, 1)$.

Solution:

The first step is to find the slope of the line:

$$m = \frac{y_1 - y_0}{x_1 - x_0} = \frac{1 - 3}{2 - (-2)} = \frac{-2}{4} = \frac{-1}{2}$$

Now we pick either of the two points and apply the point-slope form. The point $(2, 1)$ is slightly easier to work with so:

$$y - y_0 = m(x - x_0) \qquad \text{Write the point-slope form}$$
$$y - 1 = -\frac{1}{2}(x - 2) \qquad \text{Put in the numbers}$$
$$y - 1 = -\frac{1}{2}x + 1 \qquad \text{Distribute the slope}$$
$$y = -\frac{1}{2}x + 1 + 1 \qquad \text{Add 1 to both sides}$$
$$y = -\frac{1}{2}x + 2 \qquad \text{Simplify the equation}$$

So we see that the line (in slope-intercept form) is $y = -\frac{1}{2}x + 2$ whose graph looks like this:

The line $y = -\frac{1}{2}x + 2$

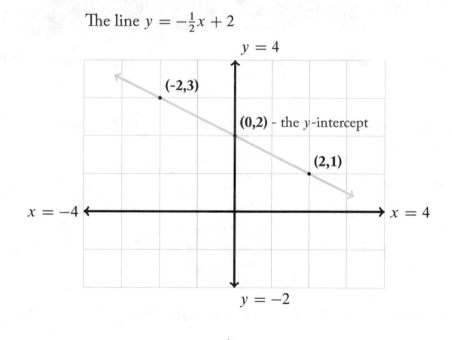

$\Diamond$

The formula for a slope had a condition: $x_0 \neq x_1$. We now need to deal with this case. When the x-coordinates of the points are the same, we get a **vertical line**.

Knowledge Box 1.5

The vertical line through the points (x_0, y_0) and (x_0, y_1) is the line

$$x = x_0$$

Unlike the other lines we have dealt with, it is not a function. In spite of that it is still a line. It contains the points: $(x_0, anything)$.

1.2.2 PARALLEL AND ORTHOGONAL LINES

Look at the lines in Figure 1.3. These three lines all have the same slope, and they are parallel to one another. It turns out that these two things always go together.

Knowledge Box 1.6

Parallel lines have the same slope.

Example 1.20 Find the line parallel to $y = 3x - 1$ that passes through the point (2,2).

Solution:

Since the line is parallel to $y = 3x - 1$, which has a slope of 3, we know that the slope of the new line is $m = 3$.

Apply the point-slope formula:

$$y - 2 = 3(x - 2)$$
$$y - 2 = 3x - 6$$
$$y = 3x - 6 + 2$$
$$y = 3x - 4$$

and we have that the line parallel to $y = 3x - 1$ that passes through (2,2) is $y = 3x - 4$.

$\Diamond$

Figure 1.4 shows two lines that intersect at right angles. We say that lines that intersect at right angles are **orthogonal**.
There is a relationship between the slopes of any two orthogonal lines.

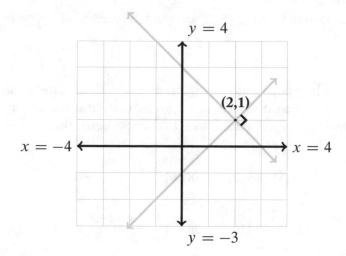

Figure 1.4: Two lines, $y = x - 1$ and $y = -x + 3$, that intersect at right angles.

Knowledge Box 1.7

Suppose that two lines intersect at right angles.
If the slope of the first line is m, then the slope of the second is:

$$-\frac{1}{m}$$

In other words: orthogonal lines have negative reciprocal slopes.

Example 1.21   Find a line at right angles to the line $y = \frac{1}{2}x + 1$ that contains the point (2,2).
Solution:
Since we want the new line to be orthogonal to $y = \frac{1}{2}x + 1$, which has a slope of $\frac{1}{2}$, the new line will have a slope of $m = -\frac{1}{1/2} = -2$. Again, we use the point-slope formula to build the line.

$$y - 2 = -2(x - 2)$$
$$y - 2 = -2x + 4$$
$$y = -2x + 6$$

Make your own graph of these two lines to see if they are, actually, at right angles.

◊

Example 1.22   Are the points (1,1), (3,3), and (6,0) the vertices of a right triangle?

Solution:

These three points can be grouped into three pairs of points. Each of these pairs of points define a line. If the three points are the vertices of a right triangle, then two of the lines form a right angle and, thus, have negative reciprocal slopes. Let's compute the three slopes of the three lines:

$$(1, 1) \text{ and } (3, 3) : \; m_1 = \frac{3-1}{3-1} = \frac{2}{2} = 1$$

$$(3, 3) \text{ and } (6, 0) : \; m_2 = \frac{0-3}{6-3} = \frac{-3}{3} = -1$$

$$(1, 1) \text{ and } (6, 0) : \; m_3 = \frac{0-1}{6-1} = \frac{-1}{5}$$

Notice that $-\dfrac{1}{m_1} = -1 = m_2$, and so two of the lines are at right angles—which means the three points are the vertices of a right triangle.

PROBLEMS

Problem 1.23   Graph the lines with the following slopes and intercepts.

1. $m = -\frac{7}{3}, b = 2$

2. $m = -1, b = 5$

3. $m = -\frac{7}{2}, b = -1$

4. $m = -2, b = -2$

5. $m = 2, b = -4$

6. $m = -\frac{10}{3}, b = 0$

7. $m = -2, b = 5$

8. $m = -\frac{2}{3}, b = -1$

9. $m = 1, b = -1$

10. $m = -\frac{13}{5}, b = 0$

Problem 1.24   For the following pairs of points, find the slope of the line through the points.

1. (0,0) and (2,2)   5. (-1,-1) and (4,2)

2. (0,0) and (2,1)   6. (2,1) and (1,2)

3. (0,3) and (3,0)   7. (2,2) and (-3,5)

4. (1,4) and (4,2)   8. (5,0) and (2,2)

Problem 1.25   True or false: it does not matter which of the points is the first or second point when you plug them into the formula for the slope of a line through two points. Write a paragraph explaining your answer.

Problem 1.26   For each of the pairs of points in Problem 1.24, find a formula for the line in point-slope form.

Problem 1.27   For each of the pairs of points in Problem 1.24, find a formula for the line in slope-intercept form.

Problem 1.28   For the lines defined in Problem 1.23 parts 1–6, find a parallel line that passes through the point (1,2). Present your answer in slope-intercept form.

Problem 1.29   For each of the following triples of points, find a line parallel to the line through the first two points that contains the third.

1. (0,1) and (2,3) and (1,5)   4. (3,6) and (1,4) and (3,1)

2. (-1,2) and (4,1) and (0,0)   5. (2,2) and (8,3) and (4,-5)

3. (4,0) and (2,7) and (2,1)   6. (5,-3) and (-2,4) and (-1,-1)

Problem 1.30   For each of the following pairs of points, find a line at right angles to the line through the points that contains the first point.

1. (0,0) and (2,2)

2. (1,3) and (3,1)

3. (-1,1) and (4,-3)

4. (2,2) and (0,0)

5. $(\frac{1}{2},\frac{3}{2})$ and (1,1)

6. (-8,4) and (3,2)

Problem 1.31   Suppose that five roses cost $10.00, and seven roses cost $12.50. Assuming the cost is given by a linear function, determine the marginal cost per rose and the base cost for operating the flower shop. Use the resulting linear function to compute the cost of a dozen roses.

Problem 1.32   Suppose that a dozen eggs cost $4.50, and eighteen eggs cost $10.50. Assuming the cost is given by a linear function, determine the marginal cost per egg and the base cost for operating the grocery store. Use the resulting linear function to compute the cost of five dozen eggs.

Problem 1.33   A ball is thrown from a rooftop in a horizontal direction. Before it hits the ground, the distance from the building is measured as well as the height of the ball above the ground as a function of time. If we neglect the effects of air resistance, which of the two quantities follows a line? Explain in a few sentences.

Problem 1.34   Water comes out of an open spigot at the bottom of a cylindrical tank. The pressure of the water still in the tank is what drives the flow. True or false: the rate of water flow is a linear function of time. Explain in a few sentences.

Problem 1.35   Which of the following triples of points are the vertices of a right triangle? Remember that you must justify your conclusions with a sentence.

1. $A = (2, 3)$, $B = (4, 7)$, and $C = (6, 1)$

2. $A = (1, 3)$, $B = (2, 6)$, and $C = (4, 2)$

3. $A = (0, 0)$, $B = (2, 4)$, and $C = (7, 1)$

4. $A = (-2, 1)$, $B = (0, 6)$, and $C = (5, 4)$

5. $A = (-3, 4)$, $B = (-1, 8)$, and $C = (1, 3)$

6. $A = (4, 4)$, $B = (1, 1)$, and $C = (0, 5)$

Problem 1.36   Consider the points (1,2) and (3,4).

1. How many points (x, y) are there that form the vertices of a right triangle together with the given points?

2. Find them.

Problem 1.37   Prove or disprove that the points (1,3), (2,0), (4,4), and (5,1) are the vertices of a square.

Problem 1.38   Construct a procedure for testing four points to see if they are the vertices of a square and demonstrate it on the points, (0,3), (2,0), (3,5), and (5,2), that do form the vertices of a square.

Problem 1.39   Find three points that, together with (0,0), are *not* the vertices of a rectangle. Give a reason your answer is correct.

Problem 1.40   True or false, and justify your answer.

- Any three distinct points not on the same line are the vertices of a triangle.

- Any four distinct points, no three of which are on the same line, are the vertices of a quadrilateral.

Problem 1.41   Suppose that t is time. Which of the following sets of points lie on a single line in the plane? In some cases plotting the points may help. You may want to consult Section 2.3 if you are unfamiliar with the sine and cosine functions.

1. Points: $(3t, t)$

2. Points: $(\cos(t), \cos(t))$

3. Points: $(\cos(t), \sin(t))$

4. Points: $(t, t^2 - t + 1)$

5. Points: $(t^2, 2t^2 + 1)$

Problem 1.42   For the points, (0,0), (1,2), (2,1), (3,5), and (2,4), there are ten possible lines that go through two of these points. What is the largest slope and the smallest slope of these lines?

Problem 1.43   Suppose we are tracking a point on a spinning disk over time. Find some computable, linear quantity that describes the point's position.

1.3   QUADRATIC EQUATIONS

After a line, the simplest type of equation is a **quadratic equation**. Its name comes from the Latin word for "square," and it is characterized by always including a squared term. When graphed, it has a shape like that shown in Figure 1.5. This shape is called a **parabola**.

Knowledge Box 1.8

A quadratic equation is an equation in the form

$$y = ax^2 + bx + c$$

where a, b, and c are constants and a $\neq$ 0.

The usual goal, when we have a quadratic equation, is to find the values that make the equation zero or **satisfy** the equation. If we have the equation $y = x^2 - 3x + 2$ and if $x = 1$, we get $1 - 3 + 2 = 0$, and so $x = 1$ satisfies the equation. Similarly, if $x = 2$, then we get $4 - 6 + 2 = 0$, and so $x = 2$ also satisfies the equation. A number that satisfies an equation is a point where the graph of the equation crosses the x-axis. A solution to an equation is also called a **root** of the equation. How can a quadratic equation have two solutions?

It turns out that a quadratic equation can have 0, 1, or 2 roots. Figure 1.6 shows how this is possible.

When the graph of the equation never crosses the x-axis, there are no roots; when it just touches the x-axis at one point, there is one root. When it crosses the x-axis twice, there are two roots. There is a formula we can use to tell how many roots a quadratic equation has. It is called the **discriminant**.

Knowledge Box 1.9

The number of roots of $ax^2 + bx + c = 0$ is determined by looking at the **discriminant:**

$$D = b^2 - 4ac$$

If $D < 0$, then the equation has no roots; if $D = 0$, then the equation has one root; if $D > 0$, then the equation has two roots.

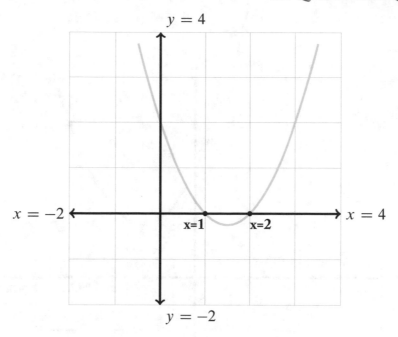

$y = 4$

$x = -2$

x=1

x=2

$x = 4$

$y = -2$

Figure 1.5: The equation $x^2 - 3x + 2 = 0$ has roots at $x = 1, 2$.

1.3.1 FACTORING A QUADRATIC

The simplest method of solving a quadratic equation is to factor it. If the equation has no roots, then it won't factor. A quadratic equation with one root is a perfect square of one term of the form $(x - a)$, possibly multiplied by a constant. When a quadratic has two roots, it factors into two different terms. Let's look at some examples of quadratics that factor:

Example 1.44

$$x^2 - 3x + 2 = (x - 1)(x - 2)$$
$$x^2 - 9 = (x - 3)(x + 3)$$
$$x^2 - 8x + 15 = (x - 5)(x - 3)$$
$$x^2 + 4x + 4 = (x + 2)(x + 2)$$
$$x^2 - x - 2 = (x - 2)(x + 1)$$
$$2x^2 + 3x + 1 = (2x + 1)(x + 1)$$

$$(Ax + B)(Cx + D) = ACx^2 + (AD + BC)x + BD$$

◊

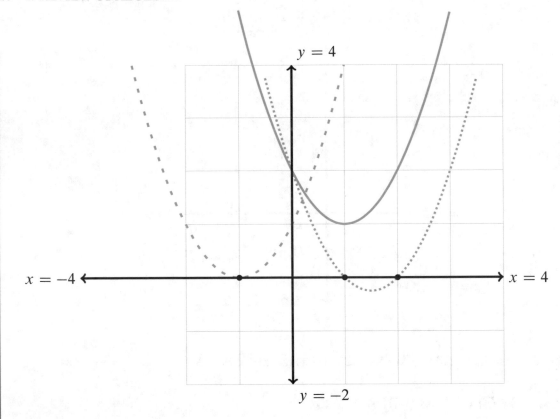

Figure 1.6: Quadratic equations with 0, 1, or 2 roots.

The last line of Example 1.44 is a **generic factorization**.

The rules for factors:

1. The product of the two constants must be the constant term of the quadratic.

2. Adding the results of multiplying the number in front of x in one term by the constant in the other term must be the number in front of x in the quadratic.

3. The product of the numbers in front of the x's must be the number in front of x^2 in the quadratic.

So, for example, $x^2 - 3x + 2 = (x - 1)(x - 2)$ uses the facts that $(-1)(-2) = 2$ and $-1 - 2 = -3$ to tell us that the factors are $(x - 1)$ and $(x - 2)$, which makes the roots 1 and 2. If $(x - a)$ is a factor of a quadratic, then a is a root of the quadratic.

Example 1.45   Factor $x^2 - 7x + 12 = 0$, but first plug in $x = 3$.

Solution:

We plug in $x = 3$ and get that $9 - 21 + 12 = 0$. So plugging in 3 makes the quadratic zero. This means:

$$x^2 - 7x + 12 = (x - 3)(x - ?)$$

But 12/3=4 so ? = 4 is a good guess. Let's check:

$$(x - 3)(x - 4) = x^2 - 3x - 4x + 12 = x^2 - 7x + 12,$$

and we have the correct factorization.

The preceding example used a trick—plugging in 3—to locate one of the factors of the quadratic. Let's make this a formal rule in Knowledge Box 1.10. In general we don't plug in 3, we plug in things that might work. Factors of the constant terms are usually good guesses—remember that they can be positive or negative.

Knowledge Box 1.10

> *Suppose that $f(x)$ is a quadratic equation and that $f(c) = 0$ for some number c. Then $(x - c)$ is a factor of $f(x)$.*

Example 1.46   Factor $x^2 - 2x - 8 = 0$.

Solution:

The number 8 factors into 1×8 and 2×4. Since $4 - 2 = 2$ and the middle term is $2x$, the factorization is probably $(x \pm 2)(x \pm 4)$. Since we have -8, one needs to be positive, and the other needs to be negative. To get a $-2x$ we need to subtract $4x$ and add $2x$. This means the factorization must be:

$$x^2 - 2x - 8 = (x - 4)(x + 2)$$

$\Diamond$

1.3.2   COMPLETING THE SQUARE

We can't always find the roots of a quadratic by factoring. Another way to find the roots is to transform the quadratic from the form:

$$ax^2 + bx + c = 0$$

into the form:

$$(x - r)^2 = s$$

Then, we can find the roots by taking the square root of both sides and doing some algebra. This is a technique called **completing the square**. Completing the square consists of forcing a complete square in the form $(x - r)^2$ and then putting the leftovers on the other side of the equation. Notice that:

$$(x - a)^2 = x^2 - 2ax + a^2$$

To complete the square we follow these steps:

1. From the equation above, we see that the constant term must be the square of half the number in front of x. We can force this by adding and subtracting the needed constant.

2. Step 1 transforms the equation into a perfect square plus a constant. Take the constant to the other side.

3. Solve by using the square root and normal algebra. Remember the $\pm$ on the square root—we will get two answers.

Example 1.47   Solve $x^2 + 2x - 1 = 0$ by completing the square.

Solution:

$$
\begin{aligned}
x^2 + 2x - 1 &= 0 & &\text{This is the original equation} \\
x^2 + 2x + (1 - 1) - 1 &= 0 & &\text{Add and subtract } \left(\tfrac{1}{2} \cdot 2\right)^2 \\
x^2 + 2x + 1 - 2 &= 0 & &\text{Regroup to get perfect square and constant} \\
(x + 1)^2 - 2 &= 0 & &\text{Make the perfect square explicit by factoring} \\
(x + 1)^2 &= 2 & &\text{Move 2 to the other side} \\
x + 1 &= \pm\sqrt{2} & &\text{Take the square root} \\
x &= -1 \pm \sqrt{2} & &\text{Move 1 to the other side}
\end{aligned}
$$

And we see the solutions to the quadratic equation are $x = -1 - \sqrt{2}$ and $x = -1 + \sqrt{2}$. This also means

$$x^2 + 2x - 1 = (x + 1 + \sqrt{2})(x + 1 - \sqrt{2}).$$

This factorization is not easy to see—which is why we need tools like completing the square.

$$\Diamond$$

There is another thing that completing the square can do.

Example 1.48 Use completing the square to show $x^2 - 4x + 7 = 0$ does not have any solutions.

Solution:

$$
\begin{array}{ll}
x^2 - 4x + 7 = 0 & \text{This is the original equation} \\
x^2 - 4x + (4 - 4) + 7 = 0 & \text{Add and subtract } \left(\frac{1}{2} \cdot -4\right)^2 \\
x^2 - 4x + 4 + 3 = 0 & \text{Regroup to get perfect square and constant} \\
(x - 2)^2 + 3 = 0 & \text{Make the perfect square explicit by factoring} \\
(x - 2)^2 = -3 & \text{Move 3 to the other side} \\
x - 2 = \pm\sqrt{-3} & \text{Take the square root. D'oh! Negative!!}
\end{array}
$$

$$\Diamond$$

Example 1.48 generated a root that was a square root of a negative. In fact, the right-hand side of the equation, just before we take the square root, is exactly the discriminant $D = b^2 - 4ac$. Knowledge Box 1.9 told us that when the discriminant is negative, there are no roots.

Another nice thing about completing the square is that it gives us the geometrically important parts of the quadratic. Quadratic equations with a positive squared term, including all the ones we've seen so far in this section, graph as parabolas opening upward. Those with a negative squared term open downward. These show up in falling object equations. The root of the perfect square that shows up when we are completing the square is the point where the quadratic turns around and heads in the other direction. If we have a downward-opening quadratic, this tells us where the highest point on the quadratic is. The next example is a physics-like example, so we are going to use time t as the independent variable instead of x.

Example 1.49 Suppose that the height in meters of a thrown ball after t seconds is given by:

$$h = 6 + 4t - t^2 \text{ m}$$

Complete the square to find when the ball is highest and at what time that greatest height occurs.

Solution:

We are going to complete the square with a small wrinkle: forcing the t^2 term positive to get the usual form for the quadratic.

$$6 + 4t - t^2 = h \qquad \text{This is the original equation}$$
$$t^2 - 4t - 6 = -h \qquad \text{Negate everything to get standard form}$$
$$t^2 - 4t + (4 - 4) - 6 = -h \qquad \text{Add and subtract } \left(\tfrac{1}{2}(-4)\right)^2$$
$$t^2 - 4t + 4 - 10 = -h \qquad \text{Regroup to get perfect square and constant}$$
$$(t - 2)^2 - 10 = -h \qquad \text{Make the perfect square explicit by factoring}$$
$$10 - (t - 2)^2 = h \qquad \text{Negate again to get } h \text{ back}$$
$$h = 10 - (t - 2)^2 \qquad \text{Neaten up}$$

We now have that h is 10 minus a squared quantity. That squared quantity is never negative and it is smallest (zero) when $t = 2$. This means the greatest height achieved by the ball is 10 meters at $t = 2$ sec, and we have our answer.

1.3.3   THE QUADRATIC EQUATION

It is possible to complete the square on the equation $ax^2 + bx + c = 0$ and get a general solution for any quadratic equation. This solution is called the **quadratic formula**.

Knowledge Box 1.11

Quadratic Formula

If $ax^2 + bx + c = 0$ *then*

$$x = \frac{-b \pm \sqrt{b^2 - 4ac}}{2a}$$

Remember that if $D = b^2 - 4ac < 0$, *then this requires us to take a square root of a negative, and so no roots exist.*

Let's try applying the quadratic equation.

Example 1.50   Find the roots of $x^2 - x - 1 = 0$.

Solution:

Apply the quadratic equation.

In this case, $a = 1, b = -1, c = -1$ so we get:

$$x = \frac{1 \pm \sqrt{1 - 4(1)(-1)}}{2 \cdot 1} = \frac{1 \pm \sqrt{5}}{2},$$

and we have the solution.

The next example shows that there is no price to using the quadratic equation when you don't need it.

Example 1.51   Find the roots of $x^2 - 4x + 3 = 0$.

Solution:

Notice that this is a problem where we could just factor the equation

$$x^2 - 4x + 3 = (x - 1)(x - 3)$$

So, we know ahead of time that $x = 1, 3$ are the solutions. Apply the quadratic equation.

In this case $a = 1, b = -4, c = 3$ so we get:

$$x = \frac{4 \pm \sqrt{16 - 4(3)(1)}}{2 \cdot 1} = \frac{4 \pm \sqrt{4}}{2} = 1, 3,$$

and we have the same solution we got by factoring.

◊

And now we do an example to show how to tell when things are not working with the quadratic equation.

Example 1.52   Find the roots of $x^2 + x + 3 = 0$.

Solution:

Apply the quadratic equation.

In this case, $a = 1$, $b = 1$, $c = 3$ so we get:

$$x = \frac{-1 \pm \sqrt{1 - 4(3)(1)}}{2 \cdot 1} = \frac{-1 \pm \sqrt{-8}}{2},$$

which is an answer containing the square root of a negative number: this is a quadratic with no roots.

$\Diamond$

The next example demonstrates that the quadratic equation results from completing the square in general. It's a bit over the top; don't worry if you can't follow it.

Example 1.53   Solve, by completing the square,

$$ax^2 + bx + c = 0.$$

Solution:

$$ax^2 + bx + c = 0 \qquad \text{This is the original equation}$$

$$x^2 + \frac{b}{a}x + \frac{c}{a} = 0 \qquad \text{Divide by } a \text{ to get a clean } x^2$$

$$x^2 + \frac{b}{a}x + \left(\frac{b^2}{4a^2} - \frac{b^2}{4a^2} \right) + \frac{c}{a} = 0 \qquad \text{Add and subtract } \frac{1}{2} \cdot \frac{b}{a} \text{ squared}$$

$$x^2 + \frac{b}{a}x + \frac{b^2}{4a^2} = \frac{b^2}{4a^2} - \frac{c}{a} \qquad \text{Move constants not in the}$$

$$\text{perfect square}$$

$$x^2 + \frac{b}{a}x + \frac{b^2}{4a^2} = \frac{b^2}{4a^2} - \frac{4ca}{4a^2}$$

Get a common denominator

$$x^2 + \frac{b}{a}x + \frac{b^2}{4a^2} = \frac{b^2 - 4ac}{4a^2}$$

Simplify the fraction

$$\left(x + \frac{b}{2a}\right)^2 = \frac{b^2 - 4ac}{4a^2}$$

Factor the perfect square

$$x + \frac{b}{2a} = \pm\frac{\sqrt{b^2 - 4ac}}{2a}$$

Take the square root of both sides

$$x = -\frac{b}{2a} \pm \frac{\sqrt{b^2 - 4ac}}{2a}$$

Take $\frac{b}{2a}$ to the other side

$$x = \frac{-b \pm \sqrt{b^2 - 4ac}}{2a}$$

Simplify: that's the quadratic formula

Since the discriminant $D = b^2 - 4ac$ is the thing we take the square root of when applying the quadratic equation, it is built right into the equation. If we're trying to take the square root of a negative number, then there is no solution; if we're taking the square root of zero, the fact that zero is its own negative means that the $\pm$ doesn't really kick in, and we get one answer.

Let's conclude the section by doing several examples. Check the values given for the examples, and see if you can compute them.

Example 1.54

Equations	Coefficients	Discriminant	Roots
$x^2 - 4x + 4 = 0$	$a = 1, b = -4, c = 4$	$D = 0$	$x = 2$
$x^2 - 4x + 3 = 0$	$a = 1, b = -4, c = 3$	$D = 4$	$x = 1, 3$
$x^2 + x + 1 = 0$	$a = 1, b = 1, c = 1$	$D = -3$	no roots
$2x^2 + 5x - 1 = 0$	$a = 2, b = 5, c = -1$	$D = 33$	$x = \dfrac{-5 \pm \sqrt{33}}{4}$
$x^2 - 7x + 4 = 0$	$a = 1, b = -7, c = 4$	$D = 33$	$x = \dfrac{7 \pm \sqrt{33}}{2}$
$x^2 + 5 = 0$	$a = 1, b = 0, c = 5$	$D = -20$	no roots

PROBLEMS

Problem 1.55   Which of the following are quadratics? Demonstrate your answer is correct. Use the definition from Knowledge Box 1.8.

1. $y = 3x + 1$

2. $y = 2x^2 + 3x - 5$

3. $y = (x - 3)(2x - 1)$

4. $y = (x + 3)^2 - 7$

5. $y = (x - 1)(x - 2)(x - 3) - (x + 1)(x + 2)(x + 3)$

6. $y = (x - 1.23)^3 + 1$

Problem 1.56   Carefully graph each of the quadratic equations in Problem 1.55.

Problem 1.57   For each of the following quadratic equations, find the discriminant D of the equation and state the number of roots the equation has.

1. $y = x^2 - 5x + 6$

2. $y = x^2 - 4x + 5$

3. $y = x^2 + 6x + 9$

4. $y = 2x^2 - x - 3$

5. $y = 2x^2 - x + 2$

6. $y = 5 - x^2$

7. $y = x^2 + x + 1$

8. $y = x^2 + x - 1$

Problem 1.58   Factor each of the following quadratics. Not all of them factor without a bit of fiddling.

1. $y = x^2 - 5x + 6$

2. $y = x^2 + 7x + 12$

3. $y = 2x^2 + 6x + 4$

4. $y = -4 + 4x - x^2$

5. $y = (x - 1)(x + 1) - 3$

6. $y = x^2 - 25$

7. $y = x^2 - 3$

8. $y = x^2 - x - 1$

Problem 1.59   For each of the following quadratics, find all the roots or give a reason there are none.

1. $y = x^2 + 5x + 6$

2. $y = x^2 + x - 6$

3. $y = x^2 - 36$

4. $y = x^2 - 20x + 91$

5. $y = x^2 + x + 1$

6. $y = x^2 + x - 1$

7. $y = x^2 + 36$

8. $y = 20x^2 + 9x + 1$

Problem 1.60   For each of following quadratics, complete the square.

1. $y = x^2 - 2x - 2$

2. $y = x^2 + 4x - 3$

3. $y = 4x^2 + 4x - 2$

4. $y = x^2 - x - \frac{3}{4}$

5. $y = x^2 + 2x + 2$

6. $y = x^2 + 6x + 10$

7. $y = x^2 + x + 1$

8. $y = x^2 + 3x + 1$

Problem 1.61   If the height of a ball at time t in meters is given by

$$h = 30 + 8t - 5t^2$$

find (without calculus) the time that ball attains its greatest height and the number of meters above the ground that the ball is at that time.

Problem 1.62   If the height of a ball at time t in meters is given by

$$h = 20 + 16t - 5t^2$$

find (without calculus) the time that ball attains its greatest height and the number of meters above the ground that the ball is at that time.

Problem 1.63   If the height of a ball at time t in meters is given by

$$h = 24 + 4t - 5t^2$$

find (without calculus) the time that ball attains its greatest height and the number of meters above the ground that the ball is at that time.

Problem 1.64   Find a quadratic equation $y = ax^2 + bx + c$ with a, b, and c all whole numbers that has two roots, neither of which is a whole number divisor of c.

Problem 1.65   Find all values of q for which $y = x^2 - qx + 2$ has two roots.

Problem 1.66   Find all values of q for which $y = x^2 - 3x + q$ has two roots.

Problem 1.67   Find a quadratic $y = ax^2 + bx + c$ so that the equation has only one root at $x = 2$ and so that, when you plug in 3 for x, $y = 2$.

Problem 1.68   Find two quadratic equations whose graphs have no points in common. Explain why your solution is correct.

Problem 1.69   Find a quadratic equation that passes through the points (-1,0), (0,1), and (1,0).

Problem 1.70   Find a quadratic equation that passes through the points (0,3), (1,2), and (2,3).

Problem 1.71   Find a quadratic equation that passes through the points (2,-1), (3,4), and (4,9).

Problem 1.72   Suppose we have two points with distinct x coordinates. How many quadratic equations have graphs that include those two points?

Problem 1.73   Find and graph a quadratic equation that has roots at $x = 2, -3$.

Problem 1.74   Find a quadratic equation with a root at $x = -2$ that also passes through the point (1,1).

Problem 1.75   Find a quadratic equation with a root at $x = 3$ that also passes through the point (-1,2).

Problem 1.76   Find a quadratic equation with a root at $x = 4$ that also passes through the point (1,6).

Problem 1.77   Suppose we have three points on the same line. Can a quadratic equation pass through all three of those points?

Problem 1.78   Consider sets of three points that are not on the same line. What added conditions are needed to permit a quadratic equation to pass through all of the points?

Problem 1.79   Suppose that $y = ax^2 + bx + c$ is a quadratic equation with no roots and that, when you plug in 2 for x, $y = 4$. What can we deduce about a?

Problem 1.80   A quadratic equation is said to be a perfect square if it has the form $f(x) = (x - a)^2$ for some constant a. What is the discriminant of a perfect square?

Problem 1.81   Find three different quadratic equations with roots 2 and -2.

Problem 1.82   Suppose we have a quadratic equation with two roots. Is there another quadratic equation whose graph has exactly the same shape but that has a root only at zero. Either explain why not or find an example for $y = x^2 - 3x + 2$.

Problem 1.83   Suppose that the graph of two quadratic equations enclose a finite area. What can you deduce about the equations from this fact?

1.4 FUNCTIONS

A **function** is a way of assigning input numbers to output numbers so that each input number is assigned to one and only one output number. When we write (x, y) this means an **ordered pair** of numbers with x first and y second. Formally:

Knowledge Box 1.12

*A **function** is a set S of ordered pairs (x, y) in which no x value appears twice.*

There is a problem with this definition—we haven't said what a set is yet. For now, a **set** is a collection of objects in which no objects appear twice.

When we are graphing a function there is a simpler way of capturing the notion of a function.

Knowledge Box 1.13

The graph of a function intersects any vertical line at most once.

If you think about it, the vertical-line based definition agrees completely with the ordered-pair definition. Why have the ordered pair definition then? It will turn out to apply in all sorts of places where the vertical line definition doesn't even make sense; stay tuned and alert.

Example 1.84   Look at the two graphs below. The one on the left is the graph of $x^2 + y^2 = 1$, while the one on the right is the graph of $y = x^2 - 1$.

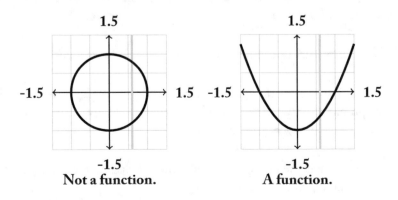

There are several places where a vertical line (we included one) hits the circle twice—it is clearly not a function. The parabola that is the graph of $y = x^2 - 1$, however, intersects a vertical line once and only once—even in the parts we cannot see.

Once we have the idea of functions, we adopt a new way of writing formulas when they happen to be functions.

Knowledge Box 1.14

Notation for functions

When we have y equal to some formula, that may or may not give us a function. When it does, we use the name $f(x)$ instead of y. The symbol $f(x)$ is still a y-value, but writing it this way tells us that x is the input variable and also lets us know that this formula is a function.

Example 1.85   Where before we might have said

$$y = x^2 + 3x + 5,$$

the fact that any quadratic in this form is a function lets us say

$$f(x) = x^2 + 3x + 5.$$

Functions are pretty abstract right now, but they keep showing up later, so we wanted to get the basic concept in front of you now. As we develop new structures from now on, we will be careful to tell you which ones are or are not functions. We are a bit behind on that promise so:

Knowledge Box 1.15

Any line $y = mx + b$ (not a vertical line) is a function. A quadratic equation $y = ax^2 + bx + c$ is also a function—each x is assigned to only one y by the formula.

One reason to use formulas to specify functions is that, as long as there is no ambiguity in the statement of the function, the formula maps each x to a single y or possibly to nothing at all if something impossible (e.g., dividing by zero) happens. The resulting collection of pairs (x, y) obeys the definition of function.

For now, circles are our big example of something that's not a function. This means that there is no way to write $f(x) = formula$ so that all the points on the circle show up as $(x, f(x))$ for some choice of x. No matter what the formula is, there will be points we cannot compute with it—unless the formula somehow violates the vertical line rule, e.g., by incorporating $\pm$, which isn't permitted.

1.4.1   DOMAIN AND RANGE

One thing we need to worry about is a formal way to define what numbers can go into or come out of a function. Negative numbers don't have a square root, for example.

Knowledge Box 1.16

*The **domain** of a function is the set of numbers that we can put into the function without a problem. Problems occur when dividing by zero or taking the square root of a negative number.*

Example 1.86   What is the domain of the function

$$f(x) = \frac{1}{x - 1}?$$

Solution:

We can take 1 divided by any number *except* zero. This means that any number can be plugged into this function except 1, which makes the thing we are dividing by zero. This means the answer is $x \neq 1$.

◊

Knowledge Box 1.17

*The **range** of a function is the set of numbers that can come out of the function.*

Example 1.87   What is the range of the function

$$f(x) = x^2?$$

Solution:

When we square a negative number, the result is positive. Since any positive number has a square root, every positive number is the result of squaring another number. We also have that $0^2 = 0$. That means any number that is not negative can come out of $f(x) = x^2$, and we see that the range is *all non-negative numbers*.

The domains and ranges of functions are usually intervals of numbers or individual numbers. At this point we want to introduce some notation that permits us to efficiently specify domains and ranges.

Definition 1.2 *If we have two sets S and T, then the* **union** *of S and T is the set of all objects that are in either set. It is written:*

$$S \bigcup T$$

Definition 1.3 *If we have two sets S and T, then the* **intersection** *of S and T is the set of all objects that are in both sets. It is written:*

$$S \bigcap T$$

Definition 1.4 *An interval from the number a to the number b with a < b has four forms. It can include both, neither, or one of a and b. We use square brackets [ ] to denote inclusion and parenthesis ( ) to denote exclusion.*

Knowledge Box 1.18

Standard and compact interval notation for intervals

$$a < x < b \quad (a, b)$$
$$a < x \leq b \quad (a, b]$$
$$a \leq x < b \quad [a, b)$$
$$a \leq x \leq b \quad [a, b]$$

Example 1.88   What are the domain and range of the function

$$f(x) = \sqrt{x^2 - 4}?$$

Solution:

To find the domain, we need to check for x values that are impossible. The only impossible thing that happens in this function is that, for some values of x, the values inside the square root function are negative. Solve:

$$x^2 - 4 = 0$$
$$(x - 2)(x + 2) = 0$$
$$x = \pm 2$$

We see that the function has roots at ± 2. Since the quadratic has a positive x^2 term, this means it opens upward, and so it is negative *between* the roots. This means the function is undefined there. So, the domain of the function is $-\infty < x \leq -2$ together with $2 \leq x < \infty$ or, in the more compact notation:

$$(-\infty, -2] \cup [2, \infty)$$

Now we need to find the range. No negative numbers come out of the square root, but zero does (at ± 2). The function $x^2 - 4$ goes to infinity. Since every positive number is the square root of some other positive number, we see all possible positive values will occur for some output of $x^2 - 4$. So, the range of $f(x)$ is $0 \leq y < \infty$ or

$$[0, \infty)$$

$$\Diamond$$

When we get to integration, the idea of odd and even functions will be handy.

Definition 1.5 *A function is* **even** *if, for x where the function exists,*

$$f(x) = f(-x).$$

A good example of an even function is $f(x) = x^2$. **Even functions forget signs.**

Definition 1.6 *A function is* **odd** *if, for x where the function exists,*

$$f(-x) = -f(x).$$

A good example of an odd function is $f(x) = x^3$. **Odd functions remember signs.**

There is a lot more to say about domain and range, but we cannot say most of it until you learn more of the library of functions. A lot of the remaining functions in the library show up in the next three sections, so stay tuned. One important point is that some of the new functions will add to our list of impossible situations, e.g., we cannot take the logarithm of zero or negative numbers, and the tangent function does not exist for odd multiples of $\frac{\pi}{2}$. As we introduce the library of functions, we will state the domains and ranges of each function.

PROBLEMS

Problem 1.89   Below are six graphs that may or may not be functions. Based only on the part of the graph shown, which of the following appear to be the graphs of functions?

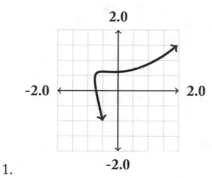

1.

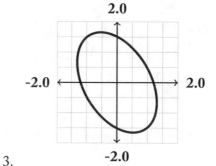

3.

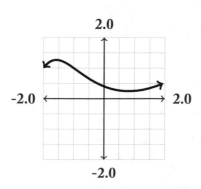

2.

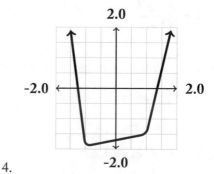

4.

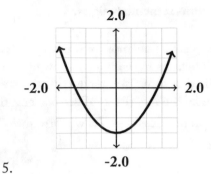

5.

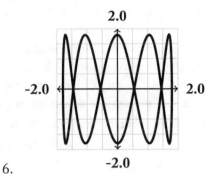

6.

Problem 1.90   Which of the following are functions? Give a reason for your answer.

1. $f(x) = 3x + 5$

2. $g(x) = x^2 + 5x + 7$

3. $h(x) = (2x - 5) + (x^2 + x + 1)$

4. $r(x) = \dfrac{x}{x^2 + 1}$

5. $s(x) = \dfrac{1 \pm \sqrt{x^2 + 3}}{x^2 + 1}$

6. $q(x) = (x - 1)(x - 2)s(x)$ where $s(x)$ is the function above

Problem 1.91   Find the domain of the following functions.

1. $f(x) = 4x - 2$

2. $g(x) = \sqrt{4x - 2}$

3. $h(x) = \dfrac{1}{4x - 2}$

4. $r(x) = \dfrac{x^2 + 1}{x^2 - 4}$

5. $s(x) = x^3 + x^2 + x + 1$

6. $q(x) = \sqrt{9 - x^2}$

7. $a(x) = (2x + 5)^3$

8. $b(x) = \sqrt[3]{2x + 5}$

Problem 1.92   Find the range of the following functions.

1. $f(x) = 5 - x$

2. $g(x) = \sqrt{5 - x}$

3. $h(x) = x^2 + 2x + 1$

4. $r(x) = 1 - x^2$

5. $s(x) = x^3$

6. $q(x) = \sqrt{25 - x^2}$

7. $a(x) = (1 - x)^3$

8. $b(x) = \sqrt[3]{x - 1}$

Problem 1.93   Which of the following sets are functions? Part (a) requires you be very careful with the definition of a function.

1. The set of points $(\cos(t), 20 \cdot \cos(t))$ where $-\infty < t < \infty$

2. All pairs (x, y) where $x^2 = y^3$

3. The points (1,2), (0,4), (-1,2), (2,5), and (-2,4)

4. The points (2,1), (4,0), (2,-1), (5,2), and (4,-2)

5. The set of points (x, y) where
$$x^2 + y^2 = 4$$
and $y \geq 0$

6. The set of points (x, y) where
$$x^2 + (y - 1)^2 = 4$$
and $y \geq 0$

Problem 1.94   For each of the functions in Problems 1.91 and 1.92 say if the functions are odd, even, or neither.

Problem 1.95   Describe carefully the largest subset(s) of a circle that are functions.

Problem 1.96   If $f(x)$ is an odd function show that $g(x) = x \cdot f(x)$ is an even function.

Problem 1.97   If $h(x)$ is an even function show that $r(x) = x \cdot h(x)$ is an odd function.

Problem 1.98   Construct an infinite set of points that has the property that the largest function that is a subset of the set contains one point.

Problem 1.99   Prove by logical argumentation that any subset of a function is a function. Remember that the definition of a function is that a function is a set of points with a special property; your argument should not touch on the graph of the function at all.

Problem 1.100   Can the graph of a function enclose a finite area? Explain your answer.

Problem 1.101   Using any method, compute the domain and range of the function

$$f(x) = \frac{ax + b}{cx + d}$$

where $a, b, c,$ and d are real numbers. You may assume that the top and bottom of the fraction do not cancel out so as to remove the variables and that $a, b \neq 0$.

Problem 1.102   Suppose that $f(x)$ and $g(x)$ are both functions whose domain is all real numbers: $(-\infty, \infty)$. If we define $h(x)$ to be equal to $f(x)$ when x is negative and to equal $g(x)$ when x is positive or zero, is $h(x)$ a function? Explain.

CHAPTER 2

The Library of Functions

This chapter continues our review of background material needed to study calculus. We start with polynomials and then review log, exponential, and trigonometric functions. If you have already had classes in these topics, you may be able to skim through this chapter, but even someone who has passed courses in these topics will benefit from reviewing the knowledge boxes.

2.1 POLYNOMIALS

We've already learned about lines, where the highest power of x is one, and quadratics, where the highest power of x is two. Polynomials are functions that let this highest power be any positive number.

Knowledge Box 2.1

*A **polynomial** is a function that is a sum of constant multiples of powers of a variable.*

Example 2.1 The following are examples of polynomials.

$$y = 3$$

$$f(x) = 3x + 4$$

$$g(x) = x^2 + 2x - 1$$

$$h(x) = x^3 + 4x - 1$$

$$k(x) = x^3 + 3x^2 + 3x + 1$$

$$q(x) = x^6 - 2$$

$$\Diamond$$

Polynomials are very well-behaved functions. They have a domain of $(-\infty, \infty)$. The range of a polynomial is a bit trickier. As we've seen, a quadratic may have a range that doesn't include everything. We need some additional terminology before we go on.

Definition 2.1 *The **standard form** of a polynomial is when it is written as a sum of constant multiples of a variable with the powers in descending order from left to right.*

The polynomials in Example 2.1 are in standard form. A example of a polynomial not in standard form is:

$$f(x) = (x - 1)(x - 2)(x^2 + 4)$$

In standard form this polynomial would be:

$$f(x) = x^4 - 3x^3 + 6x^2 - 12x + 8$$

Definition 2.2 *The **degree** of a polynomial is the highest power of x that appears in the polynomial when it is in standard form.*

The polynomials in Example 2.1 have, from top to bottom, degrees as follows: 0, 1, 2, 3, 3, 4, and 6. The degree of a polynomial may not be obvious, as we see in the next example.

Example 2.2   What is the degree of

$$f(x) = (x - 1)(x - 2)(x - 3)?$$

Solution:

To find the degree of this polynomial we have to multiply it out first.

$$\begin{aligned}
(x - 1)(x - 2)(x - 3) &= (x - 1)(x^2 - 5x + 6) \\
&= x(x^2 - 5x + 6) - 1(x^2 - 5x + 6) \\
&= x^3 - 5x^2 + 6x - x^2 + 5x - 6 \\
&= x^3 - 6x^2 + 11x - 6
\end{aligned}$$

Now, we can see that the degree of this polynomial is 3.

$\Diamond$

Actually, we don't have to multiply it out. It's possible to see that it will have a degree of three by imagining how the multiplication would come out. In fact:

Knowledge Box 2.2

If we multiply several polynomials, the degree of the product is the sum of the degrees of the polynomials we multiplied.

Definition 2.3 *The **coefficients** of a polynomial are the constants multiplied by the powers of x.*

Example 2.3   For the polynomial $f(x) = x^3 + 4x + 3$ the coefficients are 1, 4, and 3. We can be more specific: the coefficient of x^3 is 1; the coefficient of x is 4; and the constant coefficient is 3. Since there is no x^2 term, we can say that the coefficient of x^2 is 0.

Now that we know what degrees and coefficients are, we can say a little about the range of polynomial functions. These results are given here without explanation. We will revisit them in the second book, *Fast Start Integral Calculus.*

Knowledge Box 2.3

The range of a polynomial of odd degree is $(-\infty, \infty)$.

Knowledge Box 2.4

The range of a polynomial of positive even degree is:

- *(C, ∞) for some constant C if the coefficient of its highest power is positive, and*

- *$(-\infty, D)$ for some constant D if the coefficient of its highest power is negative.*

Normally "odd" and "even" cover all the possibilities, but zero is quite peculiar for an even number. $f(x) = cx^0$ is just a long way of saying $f(x) = c$, which has a very small range. Let's complete our Knowledge Box collection of possible ranges of polynomials with the following.

Knowledge Box 2.5

The range of a polynomial of degree zero is a single constant c; the function has the form $f(x) = c$.

Example 2.4

The picture on the left is the graph of a second-degree polynomial, while the one on the right is the graph of a third-degree polynomial.

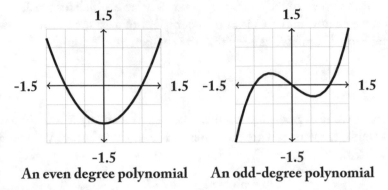

An even degree polynomial   **An odd-degree polynomial**

Check these against the Knowledge Box rules for polynomials of even and odd degree and see if they have the predicted ranges.

The next definition comes up a lot when we are trying to solve problems. We have already seen that quadratic equations can have 0, 1, or 2 roots and, in Example 1.53, found a formula for those roots. We didn't formally define what a root was then, so here it is now.

Definition 2.4 *A* **root** *of a polynomial function $f(x)$ is any number r for which $f(r) = 0$.*

If you want to think about roots geometrically, think of them as places where the graph of a function crosses the x-axis. When we are solving problems, we can sometimes write a polynomial so that the places where the polynomial is zero *are* the solutions to the problem. This will come up frequently in Chapter 4 when we are doing optimization.

Figure 2.1 shows the odd-degree polynomial from Example 2.4 with dots where its roots are.

The ranges of polynomials that we mentioned before arise from the way that a polynomial with positive degree heads toward infinity. This geometric behavior influences the roots as well. Polynomials can only change which direction they are going (up or down) a number of times that is one less than their degree. Notice a second-degree polynomial changes from down to up or from up to down exactly once. All this has the following implications for roots of polynomials.

Knowledge Box 2.6

A polynomial of odd-degree n has from one to n roots. A polynomial of positive even degree n has from zero to n roots.

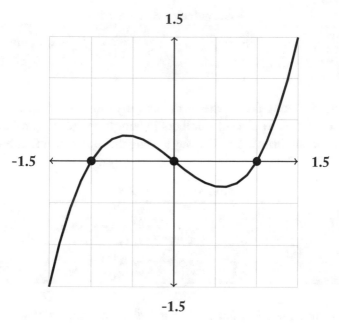

Figure 2.1: Roots of an odd-degree polynomial.

Polynomials have another interesting property: they form a **closed set** relative to addition, multiplication, and multiplication by constants. Adding or multiplying two polynomials or multiplying a polynomial by a constant results in another polynomial.

Knowledge Box 2.7

Polynomials obey the following rules:

- *A constant multiple of a polynomial is a polynomial.*

- *The sum of two polynomials is a polynomial.*

- *The product of two polynomials is a polynomial.*

Next we note that something that was true of quadratics is also true for polynomials.

Knowledge Box 2.8

Suppose that $f(x)$ is a polynomial and that $f(c) = 0$ for some number c. Then $(x - c)$ is a factor of $f(x)$.

This result is called the **root-factor theorem** for polynomials. If we're trying to factor a polynomial, one approach is to plug in numbers looking for a root (graphing the polynomial can narrow down the possibilities). Another way to state the root-factor theorem is the following.

Knowledge Box 2.9

Suppose that $f(x)$ is a polynomial and that $f(c) = 0$ for some number c. Then for some polynomial $g(x)$ we have that

$$f(x) = (x - c)g(x).$$

The second book in this series, *Fast Start Integral Calculus* contains a chapter that goes into far more detail about the properties of polynomials—something we can do once we have the tools of calculus at our fingertips.

PROBLEMS

Problem 2.5   For each of the following functions, determine if it is a polynomial. You may need to simplify the function to tell if it is a polynomial.

1. $f(x) = 1 + (x + 1) + (x^2 + x + 1) + (x^3 + x^2 + x + 1)$

2. $g(x) = (x^2 + 1)^3 + (x^2 - x + 1)^2 + 7$

3. $h(x) = \pi x + 7$

4. $r(x) = 17$

5. $s(x) = \dfrac{x}{x^2 + 1}$

6. $q(x) = x^3 + 4.1x^2 - 3.2x^2 + 4.6x - 3.8$

7. $a(x) = \dfrac{x^3 + x + 1}{x^2 + 1} - \dfrac{1}{x^2 + 1}$

8. $b(x) = (x^2 + \sqrt{x})(x^2 - \sqrt{x})$

9. $c(x) = (x^2 + \sqrt{x})^2$

10. $d(x) = \dfrac{1}{x} - \dfrac{2}{x^2} + \dfrac{3x^3 - x + 2}{x^2}$

Problem 2.6   Place each of the following polynomials into standard form.

1. $f(x) = 1 + (x + 1) + (x + 1)^2$

2. $g(x) = (x^2 + x + 1)^2$

3. $h(x) = (x^2 - 1)(x^2 + 1)(x + 2)^2$

4. $r(x) = x(x + 1)(x + 2)(x + 4)$

5. $s(x) = (x^2 + 1)(x + 1) +$
   $(x^2 + 1)(x - 2) + (x^2 + 1)^2$

6. $q(x) = x^3(x + 1)^3(x - 1)^3$

7. $a(x) = (x + 1)^3 - (x - 1)^3$

8. $b(x) = (x - 2)^3 - (x^3 - 6x^2 + 12x)$

Problem 2.7   Find the degree of each of the polynomials in Problem 2.6.

Problem 2.8   Give a simple rule for telling if a polynomial is an odd function, an even function, or neither.

Problem 2.9   Find the range of each of the following polynomials.

1. $y = 3$

2. $y = 3x + 1$

3. $y = x^2 + 6x + 12$

4. $y = x^3 + 3x^2 - 7x + 8$

5. $y = x^4 + 5x^2$

6. $y = (x + 1)(x + 2)(x + 3)(x^2 + 1)$

7. $y = (x^2 + 5)^3$

Problem 2.10   Argue convincingly that a positive whole number power of a polynomial is a polynomial.

Problem 2.11   Find a polynomial of degree 3 with one root at $x = 1$.

Problem 2.12   Find a polynomial of degree 4 with no roots at $x = 1$.

Problem 2.13   Find a polynomial of degree 3 with roots at $x = 0, \pm 3$.

Problem 2.14   Find a polynomial of degree 3 with roots at $x = 0, \pm 2$ so that $f(3) = 5$.

Problem 2.15   Prove that if two polynomials both have a root at $x = a$, then so does their sum.

Problem 2.16   Suppose that we have several polynomials. Fill in the box in the following sentence. The set of roots of the product of the polynomials is the ⬚ of the sets of roots of each of the polynomials.

Problem 2.17   True or false (and explain): if we divide two polynomials the resulting function is a polynomial.

Problem 2.18   Demonstrate that the result of dividing two polynomials can be a polynomial.

Problem 2.19   Given that each of the following graphs is the graph of a polynomial, give as much information about the degree, coefficients, and number and value of roots as you can.

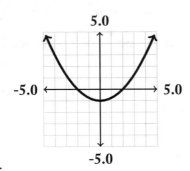

1.

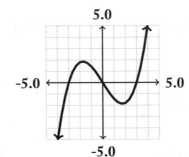

3.

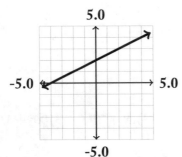

2.

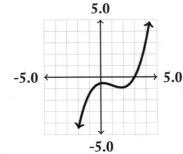

4.

5.0

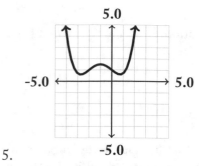

-5.0 ← → **5.0**

5.   **-5.0**

5.0

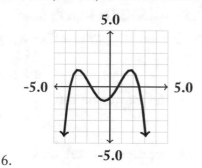

-5.0 ← → **5.0**

6.   **-5.0**

Problem 2.20   Suppose that the largest number of points that any horizontal line intersects the graph of a polynomial $f(x)$ in is m. Prove that the degree of $f(x)$ is at least m.

Problem 2.21   Give an example of a polynomial of degree 8 that has the property that the maximum number of times it intersects any horizontal line is two.

Problem 2.22   Verify that the polynomial $f(x) = x^3 - 6x^2 + 11x - 6$ has the property that $f(1) = f(2) = f(3) = 0$. Use this information and the root-factor theorem to factor $f(x)$.

Problem 2.23   Use the root factor theorem and some cleverness to factor

$$g(x) = x^4 - 625$$

Problem 2.24   If

$$h(x) = (x - 2)(x^2 + 1)(x^2 + x + 1)$$

then how many roots does $h(x)$ have and what are they?

2.2   POWERS, LOGS, AND EXPONENTIALS

This section deals with a very important category of functions: logarithms and exponential functions, as well as the related algebra of powers. We link logs and exponentials together because each can undo what the other does, like square and square root. As with square root, there is also a concern with negative numbers when computing logs. Logarithms don't exist at zero, while square roots do, so there is a difference. We begin with the algebra of powers and their corresponding roots.

2.2.1   POWERS AND ROOTS

The simplest notion of a power is that of repeated multiplication. If we multiply a number a by itself m times, we get

$$a \times a \times \cdots \times a = a^m$$

We also adopt the convention that

$$\frac{1}{a} = a^{-1}$$

which means that the reciprocal of a number to a power is that number to the negative of the power. There are a number of rules about how powers interact, summarized in the following Knowledge Box.

Knowledge Box 2.10

Algebraic Rules for Powers

- $a^{-n} = \dfrac{1}{a^n}$
- $\dfrac{a^n}{a^m} = a^{n-m}$
- $a^n \times a^m = a^{n+m}$
- $(a^n)^m = a^{n \times m}$

Example 2.25   This example showcases the rules for powers on a simple problem.

$$\left(2^3 \times 2^4\right)^5 = \left(2^{3+4}\right)^5$$
$$= \left(2^7\right)^5$$
$$= 2^{35}$$
$$= 34,359,738,368$$

The final step is probably not needed, as 2^{35} is the same number and much easier to read.

Definition 2.5 *For a positive whole number n we define*

$$b = \sqrt[n]{a}$$

to be any number for which $b^n = a$. Notice that, when n is even, there are two possibilities, $\pm \sqrt[n]{a}$.

Because an even power of a number must be positive, even roots only exist for non-negative numbers. Odd roots exist for any number, positive or negative. Once we have the notion of roots, it becomes possible to see that roots are actually a type of power.

Definition 2.6 *We define fractional powers in the following fashion:*

$$\sqrt[n]{a} = a^{\frac{1}{n}} \text{, and}$$

$$\sqrt[n]{a^m} = a^{\frac{m}{n}}$$

Example 2.26   Here are several equivalent ways of writing the fourth root of the third power of five:

$$\sqrt[4]{125} = \sqrt[4]{5^3} = \left(5^3\right)^{\frac{1}{4}} = 5^{3/4}$$

$\Diamond$

2.2.2   EXPONENTIALS AND LOGS

Exponential functions are functions involving powers in which the variables occur in the exponent. Logs are functions that undo exponential functions.

Knowledge Box 2.11

An **exponential function** *with base* **a>0** *is any function of the form*

$$y = a^x$$

This means to compute y we find the power x of a.

There are a number of difficulties with this definition of exponential functions. So far we only, strictly, know how to take whole or fractional powers of a constant a, but many numbers are not expressible as fractions. It's also sort of hard to understand what a^x means when a is negative. So, for now, we are going to avoid the whole issue of negatives and only take powers of positive numbers.

The graph in Figure 2.2 illustrates a number of properties of exponential functions of the form $y = a^x$.

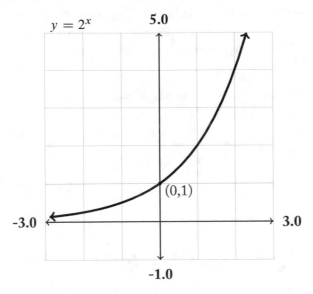

Figure 2.2: This is a graph of the function $y = 2^x$.

Knowledge Box 2.12

Properties of exponential functions

- *The domain of $y = a^x$ is $(-\infty, \infty)$.*
- *The range of $y = a^x$ is $(0, \infty)$.*
- *The graph always contains the point (0,1).*
- *For $x < 0$ and $a > 1$, a^x is a positive number smaller than $y = 1$.*
- *For $x > 0$ and $a > 1$, a^x is a positive number bigger than $y = 1$.*
- *For $y = a^{-x}$, the last two facts are reversed.*
- *If $0 < a < 1$ then the same two facts are reversed.*

The rules we just learned for powers apply to exponential functions. This means that, for example, $a^x \cdot a^y = a^{x+y}$.

Definition 2.7 *If* **c** *is the* **logarithm base b** *of a number* **a** *we write*

$$\log_b(a) = c$$

which is a different way of saying

$$b^c = a$$

The logarithm and exponential functions have the same relationship that the square and square root functions do. The technical term is that they are **inverses** of one another. Each reverses what the other does. The relationship between logs and exponentials is given in Knowledge Box 2.13.

Knowledge Box 2.13

The relationship between logs and exponentials

$$\cdot\, b^{\log_b(c)} = c \qquad\qquad \cdot \log_b(b^a) = a$$

Now that we have defined the logarithm function, we can list its algebraic properties. The fourth property is useful because it lets us compute the logarithm base *anything* once we can compute the logarithm base something.

Knowledge Box 2.14

1. $\log_b(xy) = \log_b(x) + \log_b(y)$ 3. $\log_b(x^y) = y \cdot \log_b(x)$

2. $\log_b\left(\dfrac{x}{y}\right) = \log_b(x) - \log_b(y)$ 4. $\log_c(x) = \dfrac{\log_b(x)}{\log_b(c)}$

Let's look at the graph of a logarithm function (Figure 2.3).

Now that we can see the logarithm function, let's list its domain, range, and other properties.

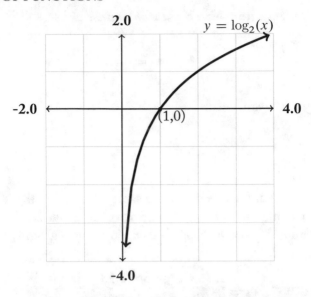

Figure 2.3: This is a graph of the function $y = \log_2(x)$.

Knowledge Box 2.15

Properties of logarithmic functions

- *The domain of $y = \log_b(x)$ is $(0, \infty)$.*

- *The range of $y = \log_b(x)$ is $(-\infty, \infty)$.*

- *The graph always contains the point (1,0).*

- *For $0 < x < 1$, $\log_b(x)$ is a negative number.*

- *For $x > 1$, $\log_b(x)$ is a positive number.*

The relationship between logs and exponentials means that we can use logs and exponentials to solve equations involving exponentials and logs. One of the rules for solving equations way back at the beginning of the chapter was "Performing the same operation to both sides, e.g., squaring or taking the square root." We add two new rules.

- We may take the logarithm of both sides of an equation.

- We may take a constant a to the power of each side of an equation.

Knowledge Box 2.16

Taking the log of both sides of an equation

A handy implication of the algebraic properties of the logarithm function is the following:

$$\text{If } b^a = c, \text{ then } a = \log_b(c).$$

This can be used to simplify and solve equations.

Example 2.27   If $2^{x+1} = 7$ find x.

Solution:

$$2^{x+1} = 7$$

$$\log_2(2^{x+1}) = \log_2(7)$$

$$x + 1 = \log_2(7)$$

$$x = \log_2(7) - 1$$

$$x \cong 1.8073549$$

$$\Diamond$$

Knowledge Box 2.17

Reversing a logarithm when solving an equation

The exponential result for solving equations with logs in them is that:
$$\text{if } \log_b(c) = a, \text{ then } c = b^a.$$

The next example shows how to deal with a logarithm in an equation by taking a constant to the power of both sides. To get rid of the logarithm, the constant has to be the base of the logarithm.

Example 2.28   If $\log_3(x^2 + x + 1) = 1.6$, find x.

Solution:

$$\log_3(x^2 + x + 1) = 1.6$$

$$3^{\log_3(x^2+x+1)} = 3^{1.6}$$

$$x^2 + x + 1 = 3^{1.6}$$

$$x^2 + x + 1 - 3^{1.6} = 0$$

At this point we have converted the problem into a quadratic and apply the quadratic equation:

$$x = \frac{-1 \pm \sqrt{1^2 - 4 \cdot 1 \cdot (1 - 3^{1.6})}}{2 \cdot 1}$$

$$x \cong 1.7471195 \text{ or } -2.7471195$$

$\Diamond$

We conclude the section on logs and exponentials with the introduction of one of the odder features of these types of functions, the **natural logarithm** based on the number e. It takes some calculus to understand why we use a nutty number like e. For now, just accept that

$$e \cong 2.7182818$$

For historical reasons there are two logarithm functions that are written without their base.

- $\ln(x)$ is shorthand for $\log_e(x)$, and

- $\log(x)$ is shorthand for $\log_{10}(x)$.

Most of the calculus of exponentials and logs is built around the twin functions $y = \ln(x)$ and $y = e^x$. These functions are mutual inverses and, in a sense we will understand later, they are the versions of the log and exponential function that arise naturally from the rest of mathematics.

PROBLEMS

Problem 2.29   For each of the following expressions, find a simplified version of the the expression as a power of a single number, as in Example 2.25, where the number was 2.

1. $\left(3^6 \cdot 3^7\right)^2$

2. $\dfrac{1}{\left(5^3 \cdot 5^4\right)^4}$

3. $\left(2^4\right)^5 \cdot \left(4 \cdot 2^3\right)$

4. $2^{\left(2^{\left(2^2\right)}\right)}$

5. $\left(\left(2^2\right)^2\right)^2$

6. $\dfrac{7 \cdot 7^2 \cdot 7^3 \cdot 7^4 \cdot 7^5}{7^6 \cdot 7^7}$

Problem 2.30   Find the power of a specified by each of these expressions.

1. $\sqrt[5]{a^{15}}$

2. $\sqrt[4]{a^3 \cdot a^5}$

3. $\dfrac{\sqrt{a^5}}{a^2 \cdot \sqrt[3]{a}}$

4. $\left(a^{1/3} \cdot a^{1/5}\right)^8$

5. $a \cdot a^{1/2} \cdot a^{1/3} \cdot a^{1/4} \cdot a^{1/5} \cdot a^{1/6}$

6. $a \cdot a^{-1/2} \cdot a^{1/3} \cdot a^{-1/4} \cdot a^{1/5}$

Problem 2.31   For each of the following functions, find the domain and range of the function. For some of these, making a plot of the function may help you find the range.

1. $f(x) = \sqrt{x^2 + 1}$

2. $g(x) = \sqrt{4 - x^2}$

3. $h(x) = \left(\sqrt[3]{x} + 1\right)^3$

4. $r(x) = \dfrac{\sqrt{1 + x}}{\sqrt{1 - x}}$

5. $s(x) = \dfrac{1}{x}$

6. $q(x) = \sqrt{x^3 - 6x^2 + 11x - 6}$

Problem 2.32   Suppose $\log_b(u) = 2$, $\log_b(v) = -1$, and $\log_b(w) = 1.2$. Compute:

1. $\log_b(u \cdot v)$

2. $\log_b(u^6 \cdot w)$

3. $\log_b\left(\dfrac{u^2 \cdot v^3}{w^{1.2}}\right)$

   4. $\log_b(b^4 \cdot w^2)$

   6. $\log_b\left(\dfrac{b^4}{u^2}\right)$

   5. $\log_b\left(\dfrac{b}{w} \cdot \dfrac{u}{v}\right)$

Problem 2.33   Solve the following for x.

   1. $2^x = 14$

   4. $2^{x^2-1} = 8$

   2. $4^x - 5 \cdot 2^x + 6 = 0$

   5. $9^x - 7 \cdot 3^x + 12 = 0$

   3. $(2^x - 8)(3^x - 9) = 0$

   6. $5^{\sqrt{x^2+1}} = 625$

Problem 2.34   For each of the following functions, find the domain and range of the function.

   1. $f(x) = 2^{x^2+1}$

   4. $r(x) = \left(\dfrac{1}{2}\right)^{-x}$

   2. $g(x) = 3^{1/x}$

   5. $s(x) = 2^x - 1$

   3. $h(x) = 1^{\sqrt{x+1}}$

   6. $q(x) = 2^x + 3^x$

Problem 2.35   Solve the following for x.

   1. $\log_5(x^2 - 6x + 8) = 1$

   4. $\ln(e^x + 1) = 2x$

   2. $2\log_3(x) = \log_3(25)$

   5. $\log_3(x + 5) = 4$

   3. $\log_2(1 - x) = 3$

   6. $\log_5(x^2 + 5x + 7) = 2$

Problem 2.36   For each of the following functions, find the domain and range of the function. Part (f) may require you to complete a square.

   1. $f(x) = \ln(x^2 + 1)$

   4. $r(x) = \sqrt{\log_2(x)^2 + 1}$

   2. $g(x) = 5 - \log_2(x)$

   5. $s(x) = \ln(2 - x)$

   3. $h(x) = \log_3(3^x + 1)$

   6. $q(x) = \log_5(x^2 + x + 1)$

Problem 2.37   Show, using the algebraic rules for exponents, that

$$\left(\frac{a}{b}\right)^{n} = \left(\frac{b}{a}\right)^{-n}$$

Problem 2.38   Show that

$$\sqrt[n]{\sqrt[m]{x}} = \sqrt[nm]{x}$$

Problem 2.39   Look at the Knowledge Boxes for properties of exponential and logarithmic functions. How are the domain and range of these two types of functions related? Explain.

Problem 2.40   What is the geometric relationship between the graphs of the logarithm and exponential functions with the same base?

Problem 2.41   Explain in what sense the equation

$$25^{x} - 4 \cdot 5^{x} + 3 = 0$$

is a quadratic. Having explained, solve it.

Problem 2.42   Suppose that $y = C \cdot a^{x}$. If the points $(0,5)$ and $(2,20)$ are on the graph, then what are C and a?

Problem 2.43   Suppose that $y = C \cdot a^{x}$. If the points $(1,6)$ and $(3,54)$ are on the graph, then what are C and a?

Problem 2.44   Suppose $y = ax + b$ is a line with positive slope. Find the domain and range of the function

$$f(x) = \sqrt{ax + b}$$

Your answer may be in terms of a and b.

Problem 2.45   Suppose $y = ax + b$ is a line with positive slope. Find the domain and range of the function

$$f(x) = \sqrt[3]{ax + b}$$

Your answer may be in terms of a and b.

Problem 2.46   Suppose $y = ax + b$ is a line with positive slope. Find the domain and range of the function

$$f(x) = \ln(ax + b)$$

Your answer may be in terms of a and b.

Problem 2.47   Suppose $y = ax + b$ is a line with positive slope. Find the domain and range of the function

$$f(x) = e^{ax+b}$$

Your answer may be in terms of a and b.

Problem 2.48   In this section it is noted that even roots, like $\sqrt{x}$, only exist for positive numbers or zero, while odd roots, like $\sqrt[3]{x}$, exist for any number. In this context discuss in a few paragraphs, trying to understand $\sqrt[\sqrt{2}]{x}$, the following question as best you can: is $\sqrt{2}$ odd, even, or neither?

Problem 2.49   Suppose

$$f(t) = A \cdot C^t$$

Compute and simplify $f(t + 1)/f(t)$.

Problem 2.50   Suppose, for positive constants C and D, that $g(t) = A \cdot C^t$ and $h(t) = B \cdot D^t$, both exponential functions. Show that

$$g(t)/h(t) = Q \cdot R^t$$

and so is also exponential. Give the conditions that determine if this function grows or shrinks as t increases.

2.3   TRIGONOMETRIC FUNCTIONS

This section does not cover trigonometry comprehensively. It assumes you already have some familiarity with the subject and focuses on the relationships between the basic trig functions, their domains and ranges, and how they relate to right triangles. A very important object for keeping track of trig functions is the **unit circle**, shown in Figure 2.4.

The unit circle, a circle of radius one centered at the origin, has the property that a ray that makes an angle θ with the x-axis intersects the unit circle at the point $(\cos(\theta), \sin(\theta))$. There are a number of angles which have sines and cosines that work out fairly evenly. These are called the **special angles**, and they are displayed on a unit circle in Knowledge Box 2.18.

Knowledge Box 2.18

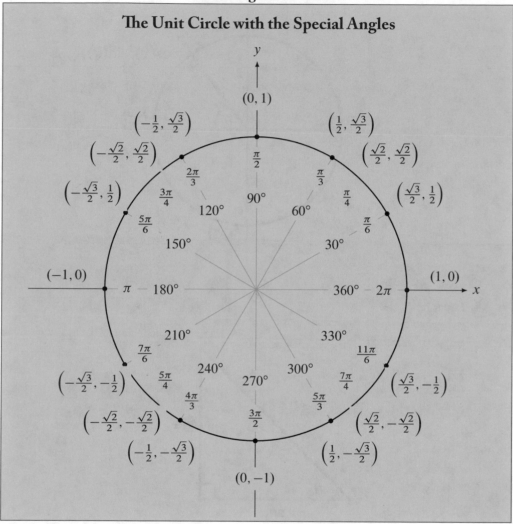

Knowledge Box 2.18 gives the angles in both degrees and radians. A full circle is 360° and 2π radians, making radians seem an odd choice. The reason for using radians—which we will do in the remainder of this text—is because they are the *natural* units of angle. A circle of radius 1 has a diameter of 2π. This means, if we use radians, that the length of an arc of a circle and the angle that subtends that arc have the same numerical value. Later on, as we develop trigonometric integrals, this will keep us from needing to perform unit conversions every time a trig function comes up in a solution.

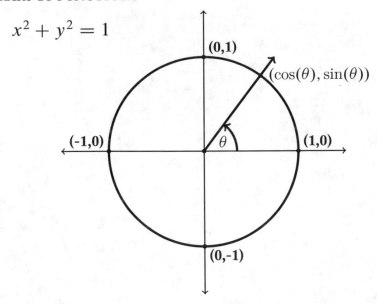

Figure 2.4: The unit circle.

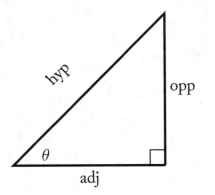

Figure 2.5: A standard right triangle showing the hypotenuse (hyp) and the sides adjacent to (adj) and opposite (opp) the angle θ.

There are six standard trigonometric functions, defined relative to the standard triangle shown in Figure 2.5. The number n in the following table represents any whole number. So $(2n + 1)$ is a way of saying "any odd whole number."

Properties and formulae for trigonometric functions.

Name	Abbrev.	Value at θ	Domain	Range
sine	sin	opp/hyp	$(-\infty, \infty)$	$[-1,1]$
cosine	cos	adj/hyp	$(-\infty, \infty)$	$[-1,1]$
tangent	tan	opp/adj	$x \neq \dfrac{2n+1}{2}\pi$	$(-\infty, \infty)$
cotangent	cot	adj/opp	$x \neq n\pi$	$(-\infty, \infty)$
secant	sec	hyp/adj	$x \neq \dfrac{2n+1}{2}\pi$	$(-\infty, -1] \cup [1, \infty)$
cosecant	csc	hyp/opp	$x \neq n\pi$	$(-\infty, -1] \cup [1, \infty)$

Knowledge Box 2.19 shows relationships between the trigonometric functions, many of them are obvious consequences of the facts in the preceding table. These are examples of **trigonometric identities**.

Knowledge Box 2.19

Some basic identities

For any angle θ, we have:

- $\tan(\theta) = \dfrac{\sin(\theta)}{\cos(\theta)}$ • $\sec(\theta) = \dfrac{1}{\cos(\theta)}$

- $\cot(\theta) = \dfrac{\cos(\theta)}{\sin(\theta)}$ • $\csc(\theta) = \dfrac{1}{\sin(\theta)}$

- $\tan(\theta) = \dfrac{1}{\cot(\theta)}$

2.3.1 GRAPHS OF THE BASIC TRIG FUNCTIONS

Look at the graphs of the sine, cosine, tangent, cotangent, secant, and cosecant functions in Figures 2.6–2.8 and check them against the domains and ranges for the functions given earlier in this section.

The trigonometric functions are **periodic**, meaning that they repeat their values regularly. This is visible in their graphs. The sine, cosine, secant, and cosecant functions repeat every 2π, while the tangent and cotangent functions repeat every π units.

The functions and co-functions have simple relationships based on sliding the graph sideways. These relationships are shown in Knowledge Box 2.20.

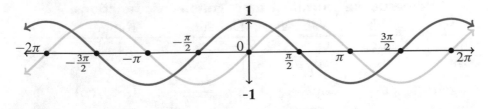

Figure 2.6: The sine (light) and cosine (dark) functions.

Knowledge Box 2.20

Periodicity identities

- $\sin(x + 2\pi) = \sin(x)$

- $\cos(x + 2\pi) = \cos(x)$

- $\sin(x) = \cos\left(x - \frac{\pi}{2}\right)$

- $\tan(x) = -\cot\left(x - \frac{\pi}{2}\right)$

- $\sec(x) = \csc\left(x + \frac{\pi}{2}\right)$

- $\cos(-x) = \cos(x)$

- $\sin(-x) = -\sin(x)$

- $\tan(x) = -\tan(x)$

- $\sin(x + \pi) = -\sin(x)$

- $\cos(x + \pi) = -\cos(x)$

- $\tan(x + \pi) = \tan(x)$

2.3.2   THEOREMS ABOUT TRIANGLES

The most basic fact about triangles is that the sum of the angles of a triangle in the plane is π radians. This means that if we know two of the angles, we can recover the third by taking π minus their sum.

If we have a right triangle with a hypotenuse of length c and legs of length a and b, then the Pythagorean theorem tells us that

$$a^2 + b^2 = c^2.$$

Homework problem 2.69 asks you to show that the fact

$$\sin^2(\theta) + \cos^2\theta = 1$$

is an instance of the Pythagorean theorem. In fact, there are several useful relations between the trigonometric functions that arise from this fact.

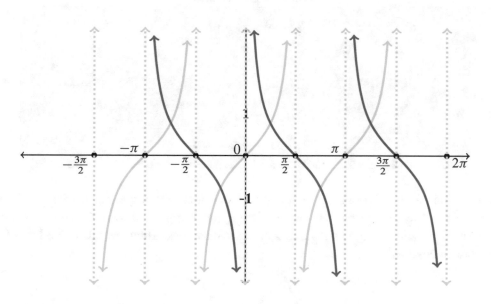

Figure 2.7: The tangent (light) and cotangent (dark) functions.

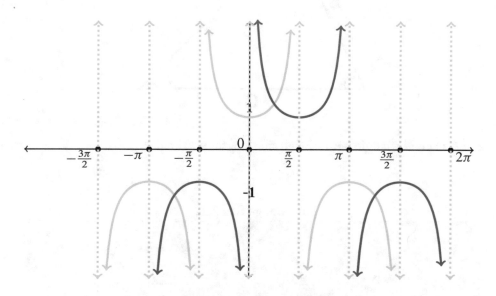

Figure 2.8: The secant (light) and cosecant (dark) functions.

Knowledge Box 2.21

The Pythagorean identities

For any angle θ, we have:

- $\sin^2(\theta) + \cos^2(\theta) = 1$
- $\tan^2(\theta) + 1 = \sec^2(\theta)$
- $1 + \cot^2(\theta) = \csc^2(\theta)$

There are some handy relationships that apply to the sides and angles of all triangles (not just right triangles). These are phrased in terms of the general triangle shown in Figure 2.9. Both of these are used to solve problems involving arbitrary triangles. They are called the **law of sines** and the **law of cosines**.

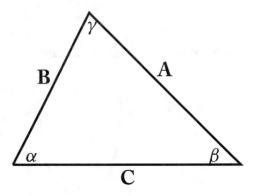

Figure 2.9: A general triangle with labeled side lengths and angles.

Knowledge Box 2.22

The law of sines

$$\frac{A}{\sin(\alpha)} = \frac{B}{\sin(\beta)} = \frac{C}{\sin(\gamma)}$$

Knowledge Box 2.23

The law of cosines

$$C^2 = A^2 + B^2 + 2AB \cdot \cos(\gamma)$$

Example 2.51   Suppose for the triangle shown in Figure 2.9 we have $A = 3$, $B = 5$, and $\beta = \dfrac{\pi}{7}$. What are α, γ, and C?

Solution:

Using the law of sines:

$$\frac{A}{\sin(\alpha)} = \frac{B}{\sin(\beta)}$$

$$\frac{3}{\sin(\alpha)} = \frac{5}{\sin(\pi/7)}$$

$$\sin(\alpha) = \frac{3\sin(\pi/7)}{5}$$

$$\alpha = \sin^{-1}\left(\frac{3\sin(\pi/7)}{5}\right)$$

$$\alpha \cong 0.273 \text{ rad}$$

For the moment, treat $\sin^{-1}$ as a button on your calculator. We will get to inverse trig functions in Section 3.3.3.

We know that $\alpha + \beta + \gamma = \pi$, so

$$\gamma = \pi - \frac{\pi}{7} - 0.273 \cong 2.42 \text{ rad.}$$

Now that we know γ, we can apply the law of sines again to get C:

$$C = \frac{\sin(\gamma) \cdot B}{\sin(\beta)} \cong 7.61$$

$\Diamond$

A **right triangle** (just in case you didn't already know) is one with a right angle in it, an angle with a value of $\pi/2$. An **isosceles triangle** is one where two of the sides have the same length. This forces it, via the law of sines, to have two equal angles as well. An **equilateral triangle** has all three sides the same and has all of its angles equal to $\dfrac{\pi}{3}$ radians. Equilateral triangles are also called **regular triangles**. Two triangles with the same three angles are **similar triangles** and, while they may be different sizes, have the same shape.

2.3.3   THE SUM, DIFFERENCE, AND DOUBLE AND HALF-ANGLE IDENTITIES

There are a large number of trig identities about the sine or cosine of the sum or difference of angles, and they are somewhat of a nightmare to memorize. From these it is not hard to derive the double and half angle identities (the last four). These last four will be very useful when we start doing integration.

Knowledge Box 2.24

Sum and difference identities; double angle identities

- $\sin(\alpha + \beta) = \sin(\alpha)\cos(\beta) + \sin(\beta)\cos(\alpha)$

- $\cos(\alpha + \beta) = \cos(\alpha)\cos(\beta) - \sin(\alpha)\sin(\beta)$

- $\sin(\alpha - \beta) = \sin(\alpha)\cos(\beta) - \sin(\beta)\cos(\alpha)$

- $\cos(\alpha - \beta) = \sin(\alpha)\sin(\beta) + \cos(\alpha)\cos(\beta)$

- $\sin(2\theta) = 2\sin(\theta)\cos(\theta)$

- $\cos(2\theta) = \cos^2(\theta) - \sin^2(\theta)$

- $\cos^2(\theta/2) = \dfrac{1 + \cos(\theta)}{2}$

- $\sin^2(\theta/2) = \dfrac{1 - \cos(\theta)}{2}$

To reiterate—this section is not a complete treatment of trigonometry. Instead it summarizes the portions of trigonometry that come up in calculus. It might seem counter intuitive that trig identities will help with calculus, but they can be used to transform expressions from things that are impossible to deal with into things that are not too hard.

2.3.4 EULER'S IDENTITY

This section contains a computational trick for quickly recovering the various sum-of-angles, difference-of-angles, and the double angle identities. This section is an enrichment section; it contains a small amount of something we normally would not cover in a first-year calculus course.

Definition 2.8 $i = \sqrt{-1}$.

The problem with the above definition is that it defines something you have been told does not exist—probably ever since you first encountered square roots. In English, i is the square root of negative one. The way you deal with this is that i is a number, albeit a funny one, with the added property that $i^2 = -1$. For the time being accept it as a notational shortcut. Once we have i available, one of the great truths of the universe, **Euler's Identity**, becomes possible to state.

Knowledge Box 2.25

Euler's Identity

$$e^{i\theta} = i\sin(\theta) + \cos(\theta)$$

In order to make use of Euler's identity we need to know a little bit about **complex numbers**.

Definition 2.9 *A* **complex number** *is a number of the form* $a + bi$ *where a and b are real numbers. We say that a is the real part of the complex number and b is the imaginary part. If* $a = 0$ *the number is an* **imaginary number**. *If* $b = 0$, *the number is a plain* **real number**.

Here are the four basic arithmetic operations for complex numbers.

1. $(a + bi) + (c + di) = (a + c) + (b + d)i$

2. $(a + bi) - (c + di) = (a - c) + (b - d)i$

3. $(a + bi) \cdot (c + di) = (ac - bd) + (ad + bc)i$

4. $\dfrac{(a + bi)}{(c + di)} = \dfrac{ac + bd}{c^2 + d^2} + \dfrac{bc - ad}{c^2 + d^2}i$

One last fact is needed before we can reap the benefits of Euler's identity. If $a + bi = c + di$, then $a = c$ and $b = d$. In English, if two complex numbers are equal, then their real parts and their imaginary parts are also equal. On to harvest results!

Example 2.52 In this example we derive the two double angle identities as the real and imaginary parts of a single expression:

$$e^{2i\theta} = e^{i\theta + i\theta}$$
$$= e^{i\theta} \cdot e^{i\theta}$$
$$i\sin(2\theta) + \cos(2\theta) = (i\sin(\theta) + \cos(\theta)) \cdot (i\sin(\theta) + \cos(\theta))$$
$$= \cos^2(\theta) - \sin^2(\theta) + i\,(2\sin(\theta)\cos(\theta))$$

Pulling out the real and imaginary parts we obtain the two double angle identities:

$$\cos(2\theta) = \cos^2(\theta) - \sin^2(\theta) \text{ (real)}$$

$$\sin(2\theta) = 2\sin(\theta)\cos(\theta) \text{ (imaginary)}$$

All the sum and difference of angle identities can be derived in a similar fashion, something we leave for the homework.

PROBLEMS

Problem 2.53   Remembering that sine and cosine are periodic with period 2π, find the exact values for the following, expressed with radicals rather than decimal numbers.

1. $\cos(5\pi/3)$

2. $\sin(7\pi/6)$

3. $\tan(\pi/3)$

4. $\sec(3\pi/4)$

5. $\cos(11\pi/4)$

6. $\cot(27\pi/4)$

7. $\sin(131\pi/4)$

8. $\sin(-11\pi/6)$

Problem 2.54   Find the set of all angles, in radians, that have a cosine of $\dfrac{\sqrt{2}}{2}$.

Problem 2.55   Find the domains of the following functions.

1. $f(x) = \sin\left(\dfrac{1}{x}\right)$

2. $g(x) = \cot(\pi x)$

3. $h(x) = \cos\left(\sqrt{x}\right)$

4. $r(x) = \sec(\cos(x))$

5. $s(x) = \csc(\sin(x))$

6. $q(x) = \sqrt{\cos(x)}$

Problem 2.56   Find the ranges of the following functions.

1. $f(x) = \tan\left(\dfrac{\pi \cdot \cos(x)}{2}\right)$

2. $g(x) = \cos\left(x^2\right)$

3. $h(x) = \csc\left(x^2\right)$

4. $r(x) = \sin\left(\dfrac{\pi}{x^2 + 1}\right)$

5. $s(x) = \tan\left(\dfrac{\pi}{x^2 + 1}\right)$

6. $q(x) = \cos\left(\dfrac{1}{x}\right)$

Problem 2.57   Give exact formulas, using radicals rather than decimals, for each of the following trig functions. Simplify your expressions as much as you can. Hint: the sum and difference of angle formulas may help.

1. $\cos\left(\dfrac{\pi}{12}\right)$

2. $\sin\left(\dfrac{7\pi}{12}\right)$

3. $\cos\left(\dfrac{5\pi}{12}\right)$

4. $\tan\left(\dfrac{5\pi}{12}\right)$

5. $\cot\left(\dfrac{\pi}{12}\right)$

6. $\sec\left(\dfrac{7\pi}{12}\right)$

Problem 2.58   Using the diagram from Figure 2.9: if $\alpha = \dfrac{\pi}{3}$, $A = 5$, and $B = 4$, what are C, β, and γ?

Problem 2.59   Using the diagram from Figure 2.9: if $\alpha = \dfrac{\pi}{3}$, $\beta = \dfrac{\pi}{4}$, and $A = 4$, what are B, C, and γ?

Problem 2.60   Using the diagram from Figure 2.9: if $\alpha = \dfrac{\pi}{5}$, $\beta = \dfrac{\pi}{5}$, and $A = 2$, what are B, C, and γ?

Problem 2.61   Show that the following are true, based on material given in this section.

1. $\sin(\theta) = \sqrt{1 - \cos^2(\theta)}$

2. $\cos(\theta) = \sqrt{1 - \sin^2(\theta)}$

3. $\tan(\theta + \tau) = \dfrac{\tan(\theta) + \tan(\tau)}{1 - \tan(\theta)\tan(\tau)}$

4. $\tan(\theta - \tau) = \dfrac{\tan(\theta) - \tan(\tau)}{1 + \tan(\theta)\tan(\tau)}$

5. $\tan(2\theta) = \dfrac{2\tan(\theta)}{1 - \tan^2(\theta)}$

6. $\tan^2(\theta) = \dfrac{1 - \cos(2\theta)}{1 + \cos(2\theta)}$

Problem 2.62   What happens to the law of cosines when the angle γ is a right angle? Refer to the definition of the law of cosines in the text.

Problem 2.63   Prove that a right isosceles triangle has, in addition to its right angle, two angles that are $\dfrac{\pi}{4}$ radians. Also show that the ratio of the side lengths of this triangle is $1 : 1 : \sqrt{2}$.

Problem 2.64   Prove that a right triangle with one of its other angles of size $\dfrac{\pi}{6}$ also has an angle of size $\dfrac{\pi}{3}$. Also show that the ratio of the side lengths of this triangle is $1 : \sqrt{3} : 2$.

Problem 2.65   Suppose that the repeated angle of an isosceles triangle has the value $\theta = \dfrac{\pi}{6}$, and that its longest side has length 2. Find the other angle and the lengths of the other sides.

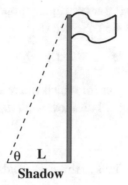

Problem 2.66   Refer to the diagram of the flag pole. If the shadow length is $L = 16m$ and the angle measures as $\theta = \dfrac{\pi}{4} \; rad$, then how tall is the flagpole?

Problem 2.67   Refer to the diagram of the flag pole. If the shadow length is $L = 6m$ and the angle measures as $\theta = 1.1 \; rad$, then how tall is the flagpole?

Problem 2.68   Refer to the diagram of the flag pole. If the shadow length is $L = 4m$ and the angle measures as $\theta = 1.4 \; rad$, then how tall is the flagpole?

Problem 2.69   Explain for which right triangle $\cos^2(\theta) + \sin^2(\theta) = 1$ is an instance of the Pythagorean theorem.

Problem 2.70   Use Euler's identity to derive the sum-of-angles identities.

Problem 2.71   Use Euler's identity to derive the difference-of-angles identities.

Problem 2.72   Use Euler's identity to derive the double angle identities.

Problem 2.73   Use Euler's identity to derive the half angle identities.

Problem 2.74   Find an expression in terms of $\sin(\theta)$ and $\cos(\theta)$ for $\sin(3\theta)$.

Problem 2.75   Find an expression in terms of $\sin(\theta)$ and $\cos(\theta)$ for $\cos(3\theta)$.

Problem 2.76   A *Pythagorean triple* is a set of three whole numbers that could be the sides of a right triangle. The most famous is (3,4,5). Notice that

$$3^2 + 4^2 = 9 + 16 = 25 = 5^2.$$

Find three additional Pythagorean triples for which the numbers do not have a common whole-number divisor bigger than 1. This would disallow (6,8,10) because these three numbers have a common factor of two.

Problem 2.77   Find the angles, in radians to three decimals, for a right triangle with side lengths 3, 4, and 5.

Problem 2.78   Classify each of the six basic trig functions as odd, even, or neither.

CHAPTER 3

Limits, Derivatives, Rules, and the Meaning of the Derivative

Traditional calculus courses begin with a detailed formal discussion of limits and continuity. This book departs from that tradition, with this chapter introducing limits only in an informal fashion so as to be able to get going with calculus. A formal discussion of limits and continuity appears in Chapter 6. The agenda for this chapter is to get you on board with a workable operational definition of limits; use this to give the formal definition of a derivative; develop the rules for taking derivatives; and end with a discussion of the physical meaning of the derivative.

3.1 LIMITS

Suppose we are given a function definition like:

$$f(x) = \frac{x^2 - 4}{x + 2}$$

Then, as long as $x \neq -2$, we can simplify as follows:

$$f(x) = \frac{x^2 - 4}{x + 2} = \frac{(x - 2)(x + 2)}{(x + 2)} = \frac{(x - 2)\cancel{(x + 2)}}{\cancel{(x + 2)}} = x - 2$$

So, this function is a line—as long as $x \neq -2$. What happens when $x = -2$? Technically, the function doesn't exist. This is where the notion of a **limit** comes in handy. If we come up with a whole string of x values and look where they are going as we approach -2, they all seem to be going toward minus 4. The key phrase here is *seem to be*, and the rigorous, precise definition of this vague phrase is the meat of Chapter 6.

For now, let's examine a tabulation of the behavior of $f(x)$ near $x = -2$.

From above		From below	
x	$f(x)$	x	$f(x)$
-1	-3	-3	-5
-1.5	-3.5	-2.5	-4.5
-1.75	-3.75	2.25	-4.25
-1.8	-3.8	-2.2	-4.2
-1.9	-3.9	-2.1	-4.1
-1.95	-3.95	-2.05	-4.05
-1.99	-3.99	-2.01	-4.01
Heading for:			
-2	-4	-2	-4

Notice that this table approaches from above (numbers larger than $x = -2$) and below (numbers smaller than $x = -2$). In a well-behaved function the approaches from above and below head for the same place, but there are functions where they don't. We call these the **limit from above** and the **limit from below**. If they agree, their joint value is the **limit of the function**. In this case the limit of the function at $x = -2$ is -4.

Definition 3.1 *We use the following symbols for the limits from above and below and the limit of a function $f(x)$ at a point $x = c$:*

$$\lim_{x \to c+} f(x) \qquad\qquad \lim_{x \to c-} f(x) \qquad\qquad \lim_{x \to c} f(x)$$

Example 3.1   What the tabulated information about $f(x) = \dfrac{x^2 - 4}{x - 2}$ suggests is that:

$$\lim_{x \to 2+} f(x) = -4 \qquad\qquad \lim_{x \to 2-} f(x) = -4 \qquad\qquad \lim_{x \to 2} f(x) = -4$$

◇

The problem with the example function $f(x) = \dfrac{x^2 - 4}{x - 2}$ is that it seems contrived. We will see shortly that functions with this sort of implausible structure arise naturally when we try to answer the simple question: **What is the tangent line to a function at a point?** This is the central question for this chapter.

Before we get there, we need the rules-of-thumb for taking limits. Suppose we are trying to compute L such that:

$$\lim_{x \to c} f(x) = L$$

We can follow these rules:

- If either of the limits from above or below don't exist, then L does not exist.

- If the limits from above and below exist but are not equal, then L does not exist.

- If the limits from above and below exist and are equal, then the limit of the function is the joint value of the upper and lower limits.

- If the function $f(x)$ is one that we can just plug into and is **continuous**, i.e., one that can be drawn without lifting our pencil (this is another Chapter 6 issue), then we can compute the limit by just plugging in.

- If the function can be turned into a well-behaved function by algebra that works everywhere but $x = c$ (like canceling $(x + 2)$ in our example), then plugging into the modified function computes the limit.

Example 3.2   Compute:
$$\lim_{x \to 2} x^3 + x^2 + x + 1$$

Solution:

Polynomials are the most well-behaved functions possible; we can always take their limits by just plugging into them. So

$$\lim_{x \to 2} x^3 + x^2 + x + 1 = 8 + 4 + 2 + 1 = 15$$

$\Diamond$

Example 3.3   Compute:

$$\lim_{x \to 1} \frac{x\,e^x - e^x}{x - 1}$$

Solution:

This function is like our original example in the sense that

$$\frac{x\,e^x - e^x}{x - 1} = \frac{e^x(x - 1)}{x - 1} = \frac{e^x(x - 1)}{x - 1} = e^x,$$

which tells us that

$$\lim_{x \to 1} \frac{x\,e^x - e^x}{x - 1} = e^1 = e.$$

$\diamond$

Example 3.4   Compute:

$$\lim_{x \to 2} \frac{x^3 - 6x^2 + 11x - 6}{x - 2}$$

Solution:

If we plug in $x = 2$ to the top of the fraction on a trial basis we get $8 - 24 + 22 - 6 = 0$. So, at $x = 2$ we have the forbidden configuration $\frac{0}{0}$. The root-factor theorem for polynomials tells us that $(x - 2)$ is a factor of $x^3 - 6x^2 + 11x - 6$. So, a little work gives us that

$$\frac{x^3 - 6x^2 + 11x - 6}{x - 2} = \frac{(x - 3)(x - 2)(x - 1)}{x - 2} = \frac{(x - 3)(x - 2)(x - 1)}{x - 2} = (x - 3)(x - 1)$$

So we have that:

$$\lim_{x \to 2} \frac{x^3 - 6x^2 + 11x - 6}{x - 2} = (2 - 3)(2 - 1) = -1$$

$\diamond$

3.1.1   SPLIT-RULE FUNCTIONS

It is sometimes desirable to have functions that obey different rules for different values of x. Retailers often offer bulk pricing discounts, for example, with different costs per unit purchased for larger and smaller numbers of units. There is a notation for this kind of function, called **split rule notation**. Suppose that $s(x)$ is a function that squares negative numbers but adds one to positive numbers and zero. Then we would say:

$$f(x) = \begin{cases} x^2 & x < 0 \\ x+1 & x \geq 0 \end{cases}$$

If we look at the graph of this function in Figure 3.1, we see that split-rule functions give us a lot of scope for creating functions that lack a limit at a point. Notice that the inequality at the change point is denoted in the graph by using a filled circle for the point that is part of the function and an empty circle for the point that is not part of the function.

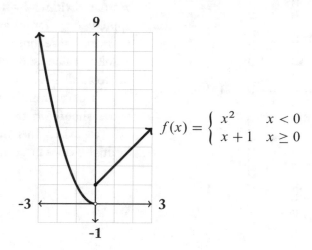

Figure 3.1: A split-rule function.

Example 3.5 Examine the following function:

$$f(x) = \begin{cases} x^2 + 1 & x < 1 \\ 3x - 1 & x \geq 1 \end{cases}$$

What is the value of $\lim\limits_{x \to 1} f(x)$?

Solution:

As we approach from below 1, the limit will be determined by the rule $x^2 + 1$, and so the limit at 1 is 2, from below. As we approach from above 1, the limit will be determined by the rule $3x - 1$, which will make the limit 2. Since the upper and lower limits both agree, the limit of the function at $x = 1$ is 2.

◊

Example 3.6   Examine the following function: $g(x) = \dfrac{1}{x-2}$ What is the value of $\lim\limits_{x \to 2} f(x)$?

Solution:

For this function we need to look at the graph, at least until we learn more:

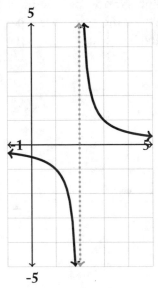

$$g(x) = \frac{1}{x-2}$$

As we approach 2 from below, the value of $g(x)$ is negative. But, since we are dividing by numbers that are approaching more and more closely to zero, those numbers get bigger in absolute value, and the function shoots off toward $-\infty$. This means the limit doesn't exist. Similarly, the limit from above shoots off toward $+\infty$ and also fails to exist. This means the desired limit also does not exist.

PROBLEMS

Problem 3.7   For each of the following limits, give a reason the limit does not exist or compute its value.

1. $\lim\limits_{x \to 3} \dfrac{x^2 - 9}{x - 3}$

2. $\lim\limits_{x \to 1} \dfrac{x^2 - 16}{x + 4}$

3. $\lim\limits_{x \to -3} \dfrac{x^3 + 6x^2 + 11x + 6}{x + 3}$

4. $\lim\limits_{x \to 0} \dfrac{e^x - 1}{e^x + 1}$

5. $\lim\limits_{x \to 0} \dfrac{e^x + 1}{e^x - 1}$

6. $\lim\limits_{x \to 1} \dfrac{2}{x^3 - 6x^2 + 11x - 6}$

7. $\lim\limits_{x \to 2} \sqrt{x - 2}$

8. $\lim\limits_{x \to 0} \ln(x)$

Problem 3.8   For each of the following limits, give a reason the limit does not exist or compute its value. Use the functions $f(x)$, $g(x)$, and $h(x)$ that follow.

$$f(x) = \begin{cases} x^3 & x \leq 1 \\ -4x + 5 & x >= 1 \end{cases}$$

$$g(x) = \begin{cases} x^2 + 2 & x < -1 \\ 3 - x^2 & x >= -1 \end{cases}$$

$$h(x) = \begin{cases} 2x + 5 & x \leq 2 \\ 9 - 2x & x >= 2 \end{cases}$$

1. $\lim\limits_{x \to -1} f(x)$

2. $\lim\limits_{x \to 1} f(x)$

3. $\lim\limits_{x \to -1} g(x)$

4. $\lim\limits_{x \to 1} g(x)$

5. $\lim\limits_{x \to -2} h(x)$

6. $\lim\limits_{x \to 2} h(x)$

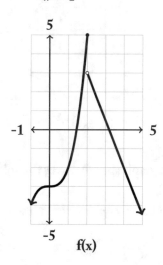

f(x)

Problem 3.9   Using $f(x)$ shown in the graph above answer the following questions.

1. What is the value of the point $x = c$ where the rules change over?

2. What is $\lim\limits_{x \to c^-} f(x)$?

3. What is $\lim\limits_{x \to c^+} f(x)$?

4. Does the limit of $f(x)$ at c exist?

Problem 3.10   For which values of c does the function

$$f(x) = \begin{cases} x^2 - 1 & x < c \\ 2x + 5 & x \geq c \end{cases}$$

have a limit at $x = c$?

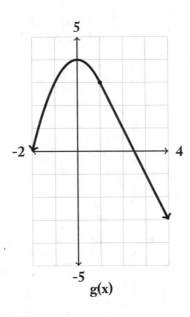

g(x)

Problem 3.11   Using $g(x)$ in the graph above answer the following questions.

1. What is the value of the point $x = c$ where the rules change over?

2. What is $\lim\limits_{x \to c^-} g(x)$?

3. What is $\lim\limits_{x \to c^+} g(x)$?

4. Does the limit of $g(x)$ at c exist?

5. What evidence is there that the function changes rules?

Problem 3.12   For which values of c does the function

$$g(x) = \begin{cases} x^2 + 7x + 1 & x < c \\ x - 8 & x \geq c \end{cases}$$

have a limit at $x = c$?

Problem 3.13 For which values of c does the function

$$h(x) = \begin{cases} x^2 + 4x & x < c \\ 3x - 2 & x \geq c \end{cases}$$

have a limit at $x = c$?

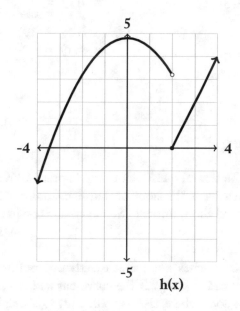

h(x)

Problem 3.14 Using $h(x)$ from the graph above answer the following questions.

1. What is the value of the point $x = c$ where the rules change over?

2. What is $\lim\limits_{x \to c^-} h(x)$?

3. What is $\lim\limits_{x \to c^+} h(x)$?

4. Does the limit of $g(x)$ at c exist?

Problem 3.15   Using the notation that an empty circle indicates a missing point on a graph, graph the following functions on the indicated interval.

1. $f(x) = \dfrac{x^2 - 1}{x + 1}$ on $[-2, 2]$

2. $g(x) = \dfrac{x^3 - 8}{x - 2}$ on $[-3, 3]$

3. $h(x) = \dfrac{x^4 - 1}{x^2 - 1}$ on $[-2, 2]$

4. $r(x) = \dfrac{x^2 - 25}{x + 5}$ on $[-8, 2]$

5. $s(x) = \dfrac{x^2 - 3}{x - \sqrt{3}}$ on $[0, 2]$

6. $q(x) = \dfrac{x^3 - 1}{x - 1}$ on $[-1, 3]$

3.2   DERIVATIVES

We mentioned earlier that the central question of this chapter is: **What is the tangent line to a function at a point?** A **derivative** is the slope of that line. In order to compute the derivative, we need to use what we learned about limits in Section 3.1. So, what is a tangent line?

A **tangent line** is a line that touches a curve at exactly one point—at least near that point. The point is called the **point of tangency**. If the curve has a complex shape, then the tangent line may intersect the curve somewhere else as well. But, in a neighborhood of the point of tangency, it brushes the curve only once. The gray line in Figure 3.2 shows a line tangent to a curve.

A **secant line** is a line through two points on a curve. Figure 3.3 shows examples of several secant lines, all of which share one point—the point of tangency in the other picture.

This picture helps us to understand why we need limits to compute slopes of tangent lines. The slopes of the secant lines are all computed based on the two points they pass through. The slope of the tangent line is based on a single point—not possible to find using the slope formula for lines. If we think of the slope of the tangent line as the limit of the slopes of secant lines from a moving point to the point of tangency, then the limit as the moving point approaches the point of tangency will be the slope of the tangent line.

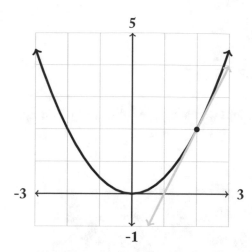

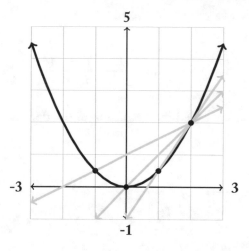

Figure 3.2: A function and a tangent line.

Figure 3.3: A function and several secant lines.

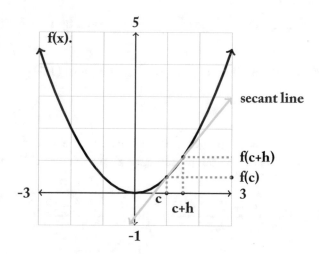

Figure 3.4: A function and a general secant line.

Suppose that the point of tangency is $(c, f(c))$, and that we examine the secant line through that point and a point "just a little" to the right—the distance to the right being h. Then, the second point on the secant line is $(c + h, f(c + h))$, giving the situation shown in the Figure 3.4. If we take the limit as $h \to 0$, then that limit should be the slope of the tangent line. Applying the formula for the slope of a line using two points, we get that the slope of the tangent line to $f(x)$ at $x = c$ is:

$$\lim_{h \to 0} \frac{f(c + h) - f(c)}{(c + h) - c} = \lim_{h \to 0} \frac{f(c + h) - f(c)}{h}$$

This formula is called the **definition of the derivative** and we have a special way of denoting it: $f'(c)$.

Knowledge Box 3.1

The slope of the tangent line to $f(x)$ at the point $(c, f(c))$ is

$$f'(c) = \lim_{h \to 0} \frac{f(c + h) - f(c)}{h}$$

In the next example, we compute the slope of a tangent line and find the formula for that tangent line.

Example 3.16   Find the tangent line to $f(x) = x^2$ at the point (1,1).

Solution:

For this problem we have $c = 1$. To get the formula for the line we need a point and a slope. We have the point (1,1) on the tangent line, so all we need to calculate is the slope.

$$\begin{aligned} \lim_{h \to 0} \frac{f(1 + h) - f(1)}{h} &= \lim_{h \to 0} \frac{(1 + h)^2 - 1^2}{h} \\ &= \lim_{h \to 0} \frac{1 + 2h + h^2 - 1}{h} \\ &= \lim_{h \to 0} \frac{2h + h^2}{h} \\ &= \lim_{h \to 0} \frac{h(2 + h)}{h} \\ &= \lim_{h \to 0} \frac{\not{h}(2 + h)}{\not{h}} \\ &= \lim_{h \to 0} 2 + h = 2 \end{aligned}$$

Notice that this limit is one that requires algebraic manipulation to resolve. We could not just plug $h = 0$ into $\dfrac{2h + h^2}{h}$ because that yields $\dfrac{0}{0}$. All tangent-slope calculations yield limits that require algebraic manipulation—explaining the emphasis on this type of limit in the previous section. We now have the point $(1, 1)$ and a slope of $m = 2$. The line is thus $y - 1 = 2(x - 1)$ or $y = 2x - 1$.

Let's conclude by graphing the function and its tangent line at $c = 1$.

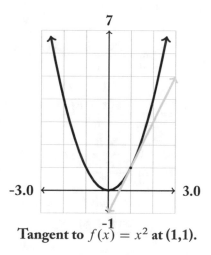

Tangent to $f(x) = x^2$ **at (1,1).**

$\Diamond$

We now know how to find the slopes of tangent lines at specific values $x = c$. It would be nice to have a general function for the derivative. We define the **general derivative** of a function as follows.

Knowledge Box 3.2

The **general derivative** *(or derivative) of* $f(x)$ *is*

$$f'(x) = \lim_{h \to 0} \frac{f(x + h) - f(x)}{h}$$

Example 3.17   Compute the derivative of $f(x) = \dfrac{1}{x}$.

Solution:

$$f'(x) = \lim_{h \to 0} \frac{f(x+h) - f(x)}{h}$$

$$= \lim_{h \to 0} \frac{\dfrac{1}{x+h} - \dfrac{1}{x}}{h}$$

$$= \lim_{h \to 0} \frac{\dfrac{x}{x(x+h)} - \dfrac{x+h}{x(x+h)}}{h}$$

$$= \lim_{h \to 0} \frac{\dfrac{x - (x+h)}{x(x+h)}}{h}$$

$$= \lim_{h \to 0} \frac{\dfrac{-h}{x(x+h)}}{h}$$

$$= \lim_{h \to 0} \frac{\dfrac{-\cancel{h}}{x(x+h)}}{\cancel{h}}$$

$$= \lim_{h \to 0} \frac{-1}{x(x+h)}$$

$$= \frac{-1}{x^2}$$

So, for $f(x) = \dfrac{1}{x}$ we have that $f'(x) = \dfrac{-1}{x^2}$.

$$\Diamond$$

The quantity

$$\frac{f(x+h) - f(x)}{h}$$

is called the **difference quotient** for $f(x)$. We can say that the derivative of a function is the limit of the difference quotient as $h \to 0$.

Example 3.18   Find the derivative of $f(x) = x^n$.

Solution:

$$f'(x) = \lim_{h \to 0} \frac{f(x+h) - f(x)}{h}$$

$$= \lim_{h \to 0} \frac{(x+h)^n - x^n}{h}$$

$$= \lim_{h \to 0} \frac{x^n + h \cdot n \cdot x^{n-1} + h^2 \cdot \text{stuff} - x^n}{h}$$

$$= \lim_{h \to 0} \frac{h \cdot n \cdot x^{n-1} + h^2 \cdot \text{stuff}}{h}$$

$$= \lim_{h \to 0} \frac{h \left(n \cdot x^{n-1} + h \cdot \text{stuff} \right)}{h}$$

$$= \lim_{h \to 0} \frac{\cancel{h} \left(n \cdot x^{n-1} + h \cdot \text{stuff} \right)}{\cancel{h}}$$

$$= \lim_{h \to 0} n \cdot x^{n-1} + h \cdot \text{stuff}$$

$$= nx^{n-1}$$

$\Diamond$

This result is our first general purpose rule for derivatives, the **power rule**.

Knowledge Box 3.3

The power rule for derivatives

If $f(x) = x^n$ then

$$f'(x) = nx^{n-1}$$

3.2.1   DERIVATIVES OF SUMS AND CONSTANT MULTIPLES

If $\lim_{x \to c} f(x) = L$ and $\lim_{x \to c} g(x) = M$ both exist, then

$$\lim_{x \to c} f(x) + g(x) = L + M.$$

Similarly, if a is a constant, then

$$\lim_{x \to c} a \cdot f(x) = a \cdot L.$$

Since derivatives are based on limits, we get two very handy rules from these facts.

Knowledge Box 3.4

Two rules for derivatives

1. The derivative of a sum is the sum of the derivatives:

$$(f(x) + g(x))' = f'(x) + g'(x)$$

2. If a is a constant, then

$$(a \cdot f(x))' = a \cdot f'(x).$$

If we plug a constant value into the difference quotient, we get zero, and the limit as $h \to 0$ of 0 is just 0. This means that the **derivative of a constant is zero**.

We already have a derivative rule for powers of x. From Section 2.1 we know that a polynomial is a sum of constant multiples of powers of x. This means that our two new rules combine with the power rule to permit us to take the derivative of any polynomial.

Example 3.19 If $f(x) = x^3 + 5x^2 + 7x + 2$ find $f'(x)$.

Solution:

Using our new rules, the power rule, and remembering that the derivative of a constant is zero, we get the following:

$$\begin{aligned}
f'(x) &= \left(x^3 + 5x^2 + 7x + 2\right)' \\
&= \left(x^3\right)' + 5\left(x^2\right)' + 7(x^1)' + 2' \\
&= 3x^2 + 5 \cdot 2x + 7 \cdot 1x^0 + 0 \\
&= 3x^2 + 10x + 7
\end{aligned}$$

and we are done.

$\diamond$

In general, to take the derivative of a polynomial in standard form, all we need to do is bring the power of each term out front, multiplying it by the existing coefficient, and subtract one from the power. So:

$$\left(x^6 + 7x^2 + 4x - 4\right)' = 6x^5 + 14x + 4$$

$$\left(3x^5 + 14x^3 - 8x^2 + 6x + 7\right)' = 15x^4 + 42x^2 - 16x + 6$$

$$\left(3x^9 - 9x^8 - 2x^7 + x^5 + 4x^4 - 7x^3 + 7x\right)' =$$

$$27x^8 - 72x^7 - 14x^6 + 5x^4 + 16x^3 - 21x^2 + 7$$

$$\left(5x^2 + 7\right)' = 10x$$

With a little practice this becomes a reflex.

PROBLEMS

Problem 3.20   Using the definition of the derivative, find $f'(c)$ for each of the following pairs of functions and constants.

1. $f(x) = x^2, c = -1$

2. $g(x) = x^3, c = 2$

3. $h(x) = \dfrac{1}{x}, c = -2$

4. $r(x) = x, c = 4$

5. $s(x) = x(x + 1), c = 0$

6. $q(x) = 3x + 7, c = 1$

Problem 3.21   For each of the following functions, give the difference quotient. In the name of providence do not attempt to simplify!

1. $f(x) = x^3$

2. $g(x) = \dfrac{1}{x^2}$

3. $h(x) = \sqrt{x}$

4. $r(x) = \cos(x)$

5. $s(x) = e^x$

6. $q(x) = \tan(x)$

Problem 3.22   Using the definition of the derivative, find the general derivative of each of the following functions.

1. $f(x) = (x + 1)^2$ 3. $h(x) = \dfrac{1}{x^2}$ 5. $s(x) = \dfrac{1}{x + 1}$

2. $g(x) = x^3$ 4. $r(x) = 17$ 6. $q(x) = x^2 + x$

Problem 3.23   For the following functions and values $x = c$, find the tangent line to the function at $x = c$ in slope-intercept form.

1. $f(x) = x^2 - 1, c = 2$ 3. $h(x) = \dfrac{1}{x}, c = 1$ 5. $s(x) = 2x^2 - 5x, c = 3$

2. $g(x) = x^3 + x^2 + x + 1, c = -1$ 4. $r(x) = \dfrac{3}{x^2}, c = 1$ 6. $q(x) = x^5 - 32, c = 2$

Problem 3.24   For $f(x) = x^2 + 1$ find the tangent lines to $f(x)$ for each of the following x-values: $\{-2, -1, 0, 1, 2\}$. Graph the tangent lines and $f(x)$ on the same set of coordinate axes.

Problem 3.25   For each of the following functions, find the derivative by any method.

1. $a(x) = 2$ 5. $h(x) = 2x^3 + 3x^2 + 7x - 11$

2. $b(x) = 115x - 234$ 6. $r(x) = 5 - x + x^2 - x^3 + x^4$

3. $f(x) = 3x^2 - 5x + 7$ 7. $s(x) = (x + 1)(x + 2)(x + 3)$

4. $g(x) = (x + 1)(x + 2)(x + 3)$ 8. $q(x) = (x + 1)^3$

Problem 3.26   Find the tangent line to $y = \sin(x)$ at $x = \dfrac{\pi}{2}$. Hint: this is a special case where you do not need the derivative to find the tangent line.

Problem 3.27 $f(x) = ax + b$ for constants a, b. Find the tangent line to $f(x)$ at $x = c$. No, you were not given an actual value for c.

Problem 3.28   For the function
$$f(x) = \sqrt{x}$$
compute the derivative using the definition of the derivative. This requires an algebra trick, but it is not too hard.

Problem 3.29   It is not too hard to compute the derivative of $f(x) = |x|$ for $x \neq 0$ because the function agrees with a line, one of $y = \pm x$, everywhere except at zero. Express and support an opinion: does the absolute value function *have* a derivative at $x = 0$?

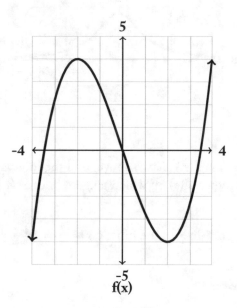

Problem 3.30   For the function $f(x)$ in the picture above, answer the following questions.

1. Find $f'(-2)$.

2. Find the tangent line to $f(x)$ at $c = 2$.

3. Find the x-values shown where the tangent line has a negative slope.

4. At which x-value does the slope of a tangent line have the largest negative value?

Problem 3.31   Find the tangent line to $f(x) = x^2 + 1$ that is parallel to the line $y = 2x - 1$.

Problem 3.32   Find the tangent line to $f(x) = x^2 + 1$ that is at right angles to the line $y = 2x - 1$.

Problem 3.33   Find a quadratic function

$$f(x) = ax^2 + bx + c$$

that has $y = 3x + 2$ as a tangent line. Demonstrate that your answer is correct.

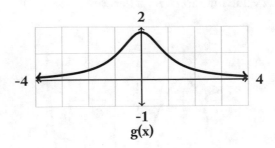

g(x)

Problem 3.34   For the function $g(x)$ shown above, find the interval(s) on which tangent lines have negative slopes and the interval(s) on which tangent lines have positive slopes.

Problem 3.35   Find all tangent lines to the function $h(x) = x^4 - 4x$ that have slope $m = 1$.

Problem 3.36   Suppose that

$$r(x) = ax^3 + bx^2 + cx + d.$$

Then what is the largest number of tangent lines that $r(x)$ can possess that have the same slope?

Problem 3.37   Find a polynomial with no roots whose first derivative has three roots.

Problem 3.38   Suppose that $s(x)$ is a quadratic polynomial. What is the geometric interpretation of the point $(x, s(c))$ where $s'(c) = 0$.

3.3   DERIVATIVES OF THE LIBRARY OF FUNCTIONS

This section is a catalog of the derivatives of the library of functions. It also introduces three new functions: the inverses of the sine, tangent, and secant functions. We can already take the derivative of polynomial functions using combinations of powers of x. It turns out that the rule for powers of the form x^n also applies for powers that are not whole numbers.

Knowledge Box 3.5

The general power rule for derivatives

If $f(x) = x^r$ for any real number r, then

$$f'(x) = rx^{r-1}$$

Example 3.39   Find the derivative of $f(x) = \sqrt{x}$.

Solution:

$$f(x) = \sqrt{x} = x^{1/2}$$

Apply the power rule and we get

$$f'(x) = \frac{1}{2}x^{1/2-1} = \frac{1}{2}x^{-1/2} = \frac{1}{2x^{1/2}} = \frac{1}{2\sqrt{x}}$$

$\Diamond$

At this point we just take off and give a whole bunch of derivative rules. It's hard to do good examples until we get to Section 3.4 where we get the rules for combining functions in various ways.

3.3.1   LOGS AND EXPONENTS

The rules for logarithm functions provide a sense of why $\ln(x)$ is called the natural logarithm. All the other logarithm functions have more complex derivative rules based on $\ln(x)$.

Knowledge Box 3.6

Derivatives of log functions

- *If $f(x) = \ln(x)$, then $f'(x) = \dfrac{1}{x}$.*

- *If $f(x) = \log_b(x)$, then $f'(x) = \dfrac{1}{x\ln(b)}$.*

The exponential function $y = e^x$ has the simplest imaginable derivative.

Knowledge Box 3.7

Derivatives of exponential functions

- *If* $f(x) = e^x$, *then* $f'(x) = e^x$.

- *If* $f(x) = a^x$, *then* $f'(x) = \ln(a) \cdot a^x$.

Example 3.40   Find the derivative of $f(x) = 3^x$.

Solution:

$$f'(x) = \ln(3) \cdot 3^x$$

$\Diamond$

3.3.2   THE TRIGONOMETRIC FUNCTIONS

The derivatives of the trigonometric functions should be committed to memory. Some patterns that help with this are as follows.

- Only the derivatives of co-functions are negative.

- A function and its co-function have derivatives that are obtained by replacing functions with co-functions.

Trigonometric derivatives

$f(x)$	$f'(x)$
$\sin(x)$	$\cos(x)$
$\cos(x)$	$-\sin(x)$
$\tan(x)$	$\sec^2(x)$
$\cot(x)$	$-\csc^2(x)$
$\sec(x)$	$\sec(x)\tan(x)$
$\csc(x)$	$-\csc(x)\cot(x)$

3.3.3   INVERSE TRIGONOMETRIC FUNCTIONS

We are already somewhat familiar with inverse functions, like the log-exponential pair and the square-square root pair, but the time has come for a formal definition.

Knowledge Box 3.9

Definition of an inverse function

A function $g(x)$ is the **inverse** *of a function $f(x)$ on an interval* $[a, b]$ *if, for all x in $[a, b]$, we have*

$$f(g(x)) = g(f(x)) = x.$$

The inverse of $f(x)$ is denoted $f^{-1}(x)$.

Example 3.41   Since $g(x) = \sqrt{x}$ only exists on the interval $[0, \infty)$, we have that $g(x) = \sqrt{x}$ is an inverse of $f(x) = x^2$ on the interval $[0, \infty)$.

◊

When $g(x)$ is an inverse of $f(x)$ on some interval, we have a special name for it. The inverse of $f(x)$ is denoted:

$$f^{-1}(x)$$

which is read "the inverse of $f(x)$." This notation is traditional but problematic because it can be confused with the negative-first power of $f(x)$, i.e., its reciprocal. Usually the meaning of a negative first power is clear from context. If in doubt, ask.

Knowledge Box 3.10

Computing inverse functions

Suppose that $y = f(x)$. If we can solve $x = f(y)$ for $y = g(x)$, then $g(x) = f^{-1}(x)$ on some interval.

Definition 3.2 *A function has a* **universal inverse** *if there is a single function that is its inverse on its entire domain.*

Example 3.42   Suppose that

$$f(x) = \frac{x + 3}{1 - x}$$

Find $f^{-1}(x)$.

Solution:

Since we have $y = \dfrac{x + 3}{1 - x}$, we solve $x = \dfrac{y + 3}{1 - y}$ for y.

$$x = \frac{y + 3}{1 - y}$$

$$x(1 - y) = y + 3$$

$$x - xy = y + 3$$

$$x - 3 = xy + y$$

$$x - 3 = y(x + 1)$$

$$\frac{x-3}{x+1} = y$$

$$y = \frac{x-3}{x+1}$$

So, we have:

$$f^{-1}(x) = \frac{x-3}{x+1}$$

Now let's check that $f(f^{-1}(x)) = x$:

$$\frac{\dfrac{x-3}{x+1} + 3}{1 - \dfrac{x-3}{x+1}} = \frac{x-3+3(x+1)}{x+1-(x-3)}$$

$$= \frac{x-3+3x+3}{x-x+1+3}$$

$$= \frac{4x}{4}$$

$$= x$$

and we have verified that the inverse is correct.

◊

We now have a firm enough grasp of inverse functions to go to work on the inverse trigonometric functions. There is an interesting feature of functions that permits them to have universal inverses—they must pass the **horizontal line test**—similar to the vertical line test for being a function. Any y-value where a horizontal line intersects the graph of a function in two places is a place where inverse values are ambiguous.

Figure 3.5 shows the horizontal line test applied to the sine function. Clearly, it fails the test. If you look back at the graphs of the other trigonometric functions in Section 2.3, you will see that all of them egregiously fail the horizontal line test. For that reason inverses are defined for only part of the domain of the trig functions. The following table gives the domain and range of each of the inverse trigonometric functions.

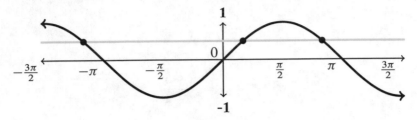

Figure 3.5: The sine function fails the horizontal line test.

Properties and formulas for trigonometric functions

Name	Abbrev.	Domain	Range
inverse sine	$\sin^{-1}$	$[-1, 1]$	$\left[-\dfrac{\pi}{2}, \dfrac{\pi}{2}\right]$
inverse cosine	$\cos^{-1}$	$[-1, 1]$	$[0, \pi]$
inverse tangent	$\tan^{-1}$	$(-\infty, \infty)$	$\left(-\dfrac{\pi}{2}, \dfrac{\pi}{2}\right)$
inverse cotangent	$\cot^{-1}$	$(-\infty, \infty)$	$(0, \pi)$
inverse secant	$\sec^{-1}$	$(-\infty, -1] \cup [1, \infty)$	$\left[0, \dfrac{\pi}{2}\right) \cup \left(\dfrac{\pi}{2}, \pi\right]$
inverse cosecant	$\csc^{-1}$	$(-\infty, -1] \cup [1, \infty)$	$\left[-\dfrac{\pi}{2}, 0\right) \cup \left(0, \dfrac{\pi}{2}\right]$

Sometimes an alternate notion is used for inverse trig functions. The prefix "arc" is added to the function name instead of the exponent -1. So, for example, $\sin^{-1}$ is written arcsin. Figures 3.6–3.8 show the graphs of the inverse trig functions, and Knowledge Box 3.11 gives the formulas for their derivatives.

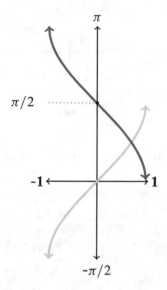

Figure 3.6: The inverse sine (light) and inverse cosine (dark) functions.

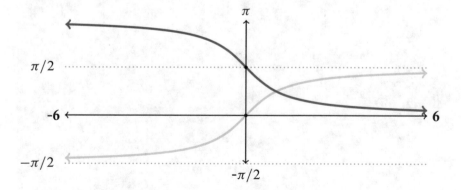

Figure 3.7: The inverse tangent (light) and inverse cotangent (dark) functions.

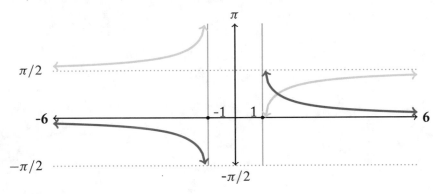

Figure 3.8: The inverse secant (light) and inverse cosecant (dark) functions.

Knowledge Box 3.11

Inverse trigonometric derivatives

$f(x)$	$f'(x)$
$\sin^{-1}(x)$	$\dfrac{1}{\sqrt{1-x^2}}$
$\cos^{-1}(x)$	$\dfrac{-1}{\sqrt{1-x^2}}$
$\tan^{-1}(x)$	$\dfrac{1}{1+x^2}$
$\cot^{-1}(x)$	$\dfrac{-1}{1+x^2}$
$\sec^{-1}(x)$	$\dfrac{1}{|x|\sqrt{x^2-1}}$
$\csc^{-1}(x)$	$\dfrac{-1}{|x|\sqrt{x^2-1}}$

PROBLEMS

Problem 3.43   Find the derivative of each of the following functions.

1. $f(x) = \sqrt[4]{x}$

2. $g(x) = x^{3.1}$

3. $h(x) = x^{\pi}$

4. $r(x) = \sin(x) + \cos(x)$

5. $s(x) = 4e^x$

6. $q(x) = \tan^{-1}(x) + \pi/2$

Problem 3.44   Find the tangent line to each of the following functions at the indicated point.

1. $f(x) = \sqrt{x}$ at $x = 1$

2. $g(x) = \sin(x)$ at $x = \pi/3$

3. $h(x) = \cos(x)$ at $x = \pi/4$

4. $r(x) = \tan^{-1}(x)$ at $x = 0$

5. $s(x) = \sin^{-1}(x)$ at $x = 0.5$

6. $q(x) = \ln(x)$ at $x = \ln(2)$

Problem 3.45   Find an inverse function for each of the following functions.

1. $f(x) = x^2 + 2x + 1$

2. $g(x) = 13x - 27$

3. $h(x) = \tan(3x + 2)$

4. $r(x) = \dfrac{2x + 1}{x - 1}$

5. $s(x) = e^{2x}$

6. $q(x) = \dfrac{1}{x}$

Problem 3.46   Find an inverse of the function $f(x) = x^2$ on the interval $(-\infty, 0]$.

Problem 3.47   Which of the following functions have universal inverses? Justify your answer.

1. $f(x) = \ln(x)$

2. $g(x) = e^x$

3. $h(x) = 2x + 5$

4. $r(x) = x^2$

5. $s(x) = x^3$

6. $q(x) = (2x + 5)^3$

Problem 3.48   Find the universal inverse of

$$f(x) = ax + b$$

when a and b are real numbers.

Problem 3.49   For which values of x does $y = \sin(x)$ have a horizontal tangent line?

Problem 3.50   Based on the information given in this section what is:

$$\lim_{x \to \infty} \tan^{-1}(x)$$

Problem 3.51   Do either of the functions $f(x) = \tan(x)$ or $g(x) = \tan^{-1}(x)$ have a universal inverse? Explain your answer.

Problem 3.52   Suppose you take the derivative of

$$y = \sin(x)$$

104 times. What do you get?

Problem 3.53   Inverses of functions exist over particular parts of their domain—a universal inverse exists everywhere in the domain. For what largest possible domains does

$$f(x) = x^2 + 4x + 4$$

have inverses. Hint: there are two answers. Find the inverses.

3.4   THE PRODUCT, QUOTIENT, RECIPROCAL, AND CHAIN RULES

In this section we learn the derivative rules that let us deal with functions built up out of other functions by both arithmetic and functional composition. Our first rule lets us take the derivative of a product of two functions. It is called the **product rule**.

Knowledge Box 3.12

The product rule

$$(f(x) \cdot g(x))' = f(x)g'(x) + f'(x)g(x)$$

Example 3.54 Find the derivative of $h(x) = x \cdot \sin(x)$.

Solution:

Apply the product rule to the functions $f(x) = x$ and $g(x) = \sin(x)$:

$$h'(x) = x \cdot (\sin(x))' + (x)' \cdot \sin(x)$$

$$= x \cdot \cos(x) + 1 \cdot \sin(x)$$

$$= x \cdot \cos(x) + \sin(x).$$

$$\Diamond$$

Example 3.55 Find the derivative of $r(x) = \ln(x) \cos(x)$.

Solution:

Apply the product rule to the functions $f(x) = \ln(x)$ and $g(x) = \cos(x)$:

$$r'(x) = \ln(x) \cdot (\cos(x))' + (\ln(x))' \cdot \cos(x)$$

$$= \ln(x) \cdot (-\sin(x)) + \frac{1}{x} \cdot \cos(x)$$

$$= \frac{\cos(x)}{x} - \ln(x) \cdot \sin(x)$$

$$\Diamond$$

The next rule is the **quotient rule** which is used to deal with the ratio of two functions.

Knowledge Box 3.13

The quotient rule

$$\left(\frac{f(x)}{g(x)} \right)' = \frac{g(x)f'(x) - f(x)g'(x)}{g^2(x)}$$

Example 3.56   Find the derivative of $h(x) = \dfrac{2x + 1}{x + 5}$.

Solution:

Apply the quotient rule to the functions $f(x) = 2x + 1$ and $g(x) = x + 5$.

$$
\begin{aligned}
h'(x) &= \frac{(x + 5)(2x + 1)' - (2x + 1)(x + 5)'}{(x + 5)^2} \\[2mm]
&= \frac{(x + 5) \cdot 2 - (2x + 1) \cdot 1}{(x + 5)^2} \\[2mm]
&= \frac{2x + 10 - (2x + 1)}{(x + 5)^2} \\[2mm]
&= \frac{9}{(x + 5)^2}
\end{aligned}
$$

In general we *do not* expand the denominator after using the quotient rule. It is often easier to deal with in factored form.

◊

Example 3.57   Find the derivative of $q(x) = \dfrac{x}{x^2 + 1}$.

Solution:

Apply the quotient rule to the functions $f(x) = x$ and $g(x) = x^2 + 1$.

$$q'(x) = \frac{(x^2 + 1) \cdot (x)' - x \cdot (x^2 + 1)'}{(x^2 + 1)^2}$$

$$= \frac{(x^2 + 1) \cdot 1 - x \cdot (2x)}{(x^2 + 1)^2}$$

$$= \frac{x^2 + 1 - 2x^2}{(x^2 + 1)^2}$$

$$= \frac{1 - x^2}{(x^2 + 1)^2}$$

◇

Example 3.58   Find the derivative of $r(x) = \dfrac{e^x}{\sin(x)}$.

Solution:

Apply the quotient rule to the functions $f(x) = e^x$ and $g(x) = \sin(x)$.

$$r'(x) = \frac{\sin(x)(e^x)' - e^x(\sin(x))'}{\sin^2(x)}$$

$$= \frac{\sin(x)e^x - e^x \cos(x)'}{\sin^2(x)}$$

$$= \frac{e^x(\sin(x) - \cos(x))}{\sin^2(x)}$$

When $f(x) = \dfrac{1}{g(x)}$, a simpler version of the quotient rule, called the **reciprocal rule**, may be used.

Knowledge Box 3.14

The reciprocal rule

$$\left(\frac{1}{f(x)}\right)' = \frac{-f'(x)}{f^2(x)}$$

Example 3.59   Find the derivative of $h(x) = \dfrac{1}{x^2 + 1}$.

Solution:

Apply the reciprocal rule to the function for which the denominator is $f(x) = x^2 + 1$.

$$h'(x) = \frac{-\left(x^2 + 1\right)'}{\left(x^2 + 1\right)^2}$$

$$= \frac{-2x}{\left(x^2 + 1\right)^2}$$

$\Diamond$

Example 3.60   Find the derivative of $r(x) = \dfrac{1}{e^x + x^2}$.

Solution:

Apply the reciprocal rule to the function for which the denominator is $f(x) = e^x + x^2$.

$$q'(x) = \frac{-\left(e^x + x^2\right)'}{\left(e^x + x^2\right)^2}$$

$$= \frac{-\left(e^x + 2x\right)}{\left(e^x + x^2\right)^2}$$

$$= -\frac{e^x + 2x}{\left(e^x + x^2\right)^2}$$

$\Diamond$

3.4.1   FUNCTIONAL COMPOSITION AND THE CHAIN RULE

The **composition** of two functions results from applying one to the other. If the functions are $f(x)$ and $g(x)$, then their composition is written $f(g(x))$. Let's look at a few examples.

Example 3.61   If $f(x) = x + 7$ and $g(x) = x^2$, then

$$f(g(x)) = x^2 + 7$$

while

$$g(f(x)) = (x + 7)^2$$

$$\Diamond$$

Example 3.62   If $f(x) = \sin(x)$ and $g(x) = e^x$, then

$$f(g(x)) = \sin(e^x)$$

while

$$g(f(x)) = e^{\sin(x)}$$

$$\Diamond$$

The order in which two functions are composed matters a lot—the results are not symmetric. The **chain rule** is used to compute the derivative of a composition of functions.

Knowledge Box 3.15

The chain rule

$$(f(g(x)))' = f'(g(x)) \cdot g'(x)$$

In the composition $f(g(x))$ we call $f(x)$ the **outer function** and $g(x)$ the **inner function**.

Example 3.63   Compute the derivative of $h(x) = e^{2x}$.

Solution:

Apply the chain rule to the functional composition for which the outer function is $f(x) = e^x$, and the inner function is $g(x) = 2x$. For these, $f'(x) = e^x$ and $g'(x) = 2$. So:

$$h'(x) = e^{2x} \cdot 2$$

$$= 2e^{2x}$$

$$\Diamond$$

Example 3.64   Compute the derivative of $q(x) = \sin\left(x^2\right)$.

Solution:

Apply the chain rule to the functional composition for which the outer function is $f(x) = \sin(x)$, and the inner function is $g(x) = x^2$. For these, $f'(x) = \cos(x)$ and $g'(x) = 2x$. So:

$$q'(x) = \cos\left(x^2\right) \cdot 2x$$

$$= 2x \cdot \cos\left(x^2\right)$$

$$\Diamond$$

The chain rule avoids a whole lot of multiplying out in some cases. Technically, we could do the following example without the chain rule, but it would be purely awful.

Example 3.65   Compute the derivative of $r(x) = \left(x^2 + x + 1\right)^7$.

Solution:

Apply the chain rule to the functional composition for which the outer function is $f(x) = x^7$, and the inner function is $g(x) = x^2 + x + 1$.

For these, $f'(x) = 7x^6$ and $g'(x) = 2x + 1$.

So:

$$r'(x) = 7\left(x^2 + x + 1\right)^6 \cdot (2x + 1)$$

Normally we don't multiply out answers like this.

◊

Example 3.66   Compute the derivative of $a(x) = \sqrt{e^x + 2}$.

Solution:

Apply the chain rule to the functional composition for which the outer function is $f(x) = \sqrt{x}$, and the inner function is $g(x) = e^x + 2$.

For these, $f'(x) = \dfrac{1}{2\sqrt{x}}$ and $g'(x) = e^x$.

So:

$$a'(x) = \frac{1}{2\sqrt{e^x + 2}} \cdot e^x$$

$$= \frac{e^x}{2\sqrt{e^x + 2}}$$

◊

Example 3.67   Compute the derivative of $b(x) = \ln(\cos(x))$.

Solution:

Apply the chain rule to the functional composition for which the outer function is $f(x) = \ln(x)$, and the inner function is $g(x) = \cos(x)$.

For these, $f'(x) = \dfrac{1}{x}$ and $g'(x) = -\sin(x)$.

So:

$$b'(x) = \frac{1}{\cos(x)} \cdot (-\sin(x))$$

$$= \frac{-\sin(x)}{\cos(x)}$$

$$= -\frac{\sin(x)}{\cos(x)}$$

$$= -\tan(x)$$

Notice how simplifying this answer requires one of the simpler trig identities.

With the product, quotient, reciprocal, and especially the chain rule, the variety of functions for which we can take the derivative is extended substantially. Practice is needed to develop a sense of which rule to use when.

PROBLEMS

Problem 3.68   Find the derivatives of the following functions with the product rule. You will need the chain rule for some of these.

1. $f(x) = x \cdot \cos(x)$

2. $g(x) = \sin(x) \cdot \cos(x)$

3. $h(x) = x^5(x+1)^6$

4. $q(x) = x^3 \cdot e^{2x}$

5. $r(x) = \tan(x) \cdot \ln(x)$

6. $s(x) = x^5 \cdot \tan^{-1}(x)$

7. $a(x) = \cos(2x)e^{3x}$

8. $b(x) = \cos(\pi x)\sin(ex)$

Problem 3.69   Find the derivatives of the following functions with the quotient rule.

1. $f(x) = \dfrac{2x + 1}{x^3 + 2}$

2. $g(x) = \dfrac{\ln(x)}{1 - x}$

3. $h(x) = \dfrac{(x + 1)^3}{(x - 1)^3}$

4. $q(x) = \dfrac{\tan(x)}{\cos(x) + 1}$

5. $r(x) = (e^x)/(1 + e^x)$

6. $s(x) = \dfrac{x^2 + 1}{x^2 + x + 1}$

7. $a(x) = \dfrac{\sin(x)}{\cos(x) + 1}$

8. $b(x) = \tan^{-1}(x)/(x^2 - 1)$

Problem 3.70   Find the derivatives of the following functions with the reciprocal rule.

1. $f(x) = \dfrac{1}{x^2 + 1}$

2. $g(x) = \dfrac{1}{\cos(x)}$

3. $h(x) = \dfrac{1}{\sqrt{2x + 1}}$

4. $q(x) = e^{-x}$

5. $r(x) = \dfrac{1}{\sin(x) + \cos(x)}$

6. $s(x) = \dfrac{1}{\ln(x)}$

Problem 3.71   Find the derivatives of the following functions with the chain rule.

1. $f(x) = \left(x^2 + 1\right)^{11}$

2. $g(x) = \cos\left(x^2 + 1\right)$

3. $h(x) = (\cos(x) + 1)^5$

4. $q(x) = \sqrt{\cos(x) + 1}$

5. $r(x) = \left(\dfrac{1}{x} + 1\right)^4$

6. $s(x) = e^{x^2 + x + 1}$

7. $a(x) = \ln\left(x^3 + 2x^2 + 7x + 5\right)$

8. $b(x) = e^{\tan(x)}$

9. $c(x) = \ln(\tan(x))$

10. $d(x) = \ln\left(e^x + 1\right)$

Problem 3.72   Find the derivatives of the following functions with the most appropriate technique(s).

1. $f(x) = \left(\dfrac{\sin(x)}{\cos(x)}\right)^5$

2. $g(x) = x \cdot \ln(\cos^2(x) + 1)$

3. $h(x) = \dfrac{(x^2 + 1)^5}{x^7 + 1}$

4. $q(x) = x^2 \cdot \ln(x^2 + 1)$

5. $r(x) = x \cdot 5^{x^2+1}$

6. $s(x) = \sin(x) \cdot \ln(x) \cdot (x^2 + 1)$

7. $a(x) = (x^3 + 1)^{121} \cdot \dfrac{x}{x + 4}$

8. $b(x) = \ln\left(\dfrac{x^3 + 1}{x^2 + 2}\right)$

Problem 3.73   Find the tangent lines to

$$f(x) = \frac{1}{x^2 + 1}$$

where $x = \pm 1$. Give your answer in point–slope form.

Problem 3.74   Find the tangent line to

$$g(x) = x\,e^{-x^2}$$

where $x = 0$. Give your answer in point–slope form.

Problem 3.75   Find the tangent line to

$$h(x) = \tan^{-1}(x/2)$$

where $x = \sqrt{3}$. Give your answer in point–slope form.

Problem 3.76   Remembering that

$$\tan(x) = \frac{\sin(x)}{\cos(x)}$$

use the quotient rule to verify the rule for the derivative of $\tan(x)$.

Problem 3.77   Remembering that

$$\cot(x) = \frac{\cos(x)}{\sin(x)}$$

use the quotient rule to verify the rule for the derivative of $\cot(x)$.

Problem 3.78   Remembering that

$$\sec(x) = \frac{1}{\cos(x)}$$

use the reciprocal rule to verify the rule for the derivative of $\sec(x)$.

Problem 3.79   Remembering that

$$\csc(x) = \frac{1}{\sin(x)}$$

use the reciprocal rule to verify the rule for the derivative of $\csc(x)$.

Problem 3.80   Derive the reciprocal rule by applying the quotient rule to

$$g(x) = \frac{1}{f(x)}$$

Problem 3.81   By using the product rule a couple times and simplifying, verify the triple product rule:

$$(f(x)g(x)h(x))' = f(x)g(x)h'(x)$$

$$+ f(x)g'(x)h(x)$$

$$+ f'(x)g(x)h(x)$$

Problem 3.82   Find the derivative of

$$f(x) = x^5(x+1)^4(x+2)^3$$

Problem 3.83   Would taking the logarithm of both sides first make Problem 3.82 any easier?

Problem 3.84   Find the derivative of
$$g(x) = x^2 \cdot \sin(x) \cdot e^x$$

Problem 3.85   Find the derivative of
$$h(x) = \sqrt{x+1} \cdot \tan(x) \cdot \ln(x)$$

Problem 3.86   By using the chain rule as many times as needed, verify the triple chain rule:
$$(f(g(h(x))))' =$$
$$f'(g(h(x))) \cdot g'(h(x)) \cdot h'(x)$$

Problem 3.87   Find the derivative of
$$f(x) = (\sin(\ln(x)))^5$$

Problem 3.88   Find the derivative of
$$g(x) = (\ln(e^x + 1))^3$$

Problem 3.89   Find the derivative of
$$g(x) = \ln\left(\sin^4\left(x^4 + 1\right)\right)$$

Problem 3.90   Prove that
$$(f(ax + b))' = a \cdot f'(ax + b)$$
for any function $f(x)$ with a derivative.

Problem 3.91   Find the derivative of $f(x) = \sin(x)\cos(x)$ with the chain rule instread of the product rule. Hint: use a trig identity.

3.5   PHYSICAL INTERPRETATION OF DERIVATIVES

So far we have gone from the limit-based definition of a derivative to knowing how to take the derivative of a pretty large number of different functions. The obvious missing piece here is the *meaning* of the derivative. The meaning of a derivative depends a lot on context, of course, but there is a general principle that covers most of what a derivative means.

Knowledge Box 3.16

What is a derivative?

If $f(x)$ measures a quantity, then $f'(x)$ is the rate at which that quantity is changing.

We have been using the geometric interpretation of the derivative: it is the slope of the tangent line to a curve at a point. A line $y = ax + b$ represents something that starts at b and adds a more per x-unit traversed; a line has a constant rate at which the quantity it is measuring changes. This explains why "$f'(x)$ is the rate of change of $f(x)$" and "$f'(x)$ is the slope of the tangent to the graph at x" are equivalent ideas.

At this point—to permit a number of innovative ways of using the derivative—we introduce a new notation that acknowledges that the derivative is a rate of change.

Knowledge Box 3.17

Differential notation

Given that $y = f(x)$, another notation for the derivative is

$$\frac{dy}{dx} = f'(x)$$

This is spoken "the differential of y with respect to x." The new symbols dy and dx are the differential of y and of x, respectively.

This notion of the derivative as a rate of change leads to a natural application in physics. We will need one more definition.

Definition 3.3 *The derivative of the derivative of a function $y = f(x)$ is called the* **second derivative** *and is denoted by*

$$\frac{d^2 y}{dx^2} = f''(x)$$

The second derivative measures the rate at which the rate of change is changing. Ouch. It also measures the **curvature** of a graph. An example of this is shown in Figure 3.9; light hash marks show the degree of downward concavity or negative curvature; dark ones document upward concavity or positive curvature.

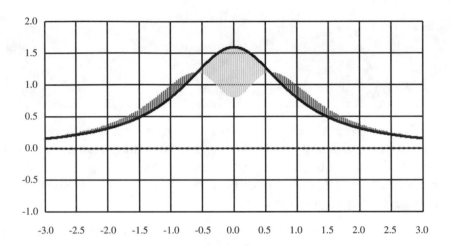

Figure 3.9: Graph of a function annotated to show curvature.

In addition to curvature, there is a straightforward interpretation of the second derivative in physics—it is **acceleration**.

Knowledge Box 3.18

Position, velocity, and acceleration

If $y = f(t)$ measures the **position** *of an object at time t, then* $\dfrac{dy}{dt} = f'(t)$ *gives its* **velocity**, *while* $\dfrac{d^2y}{dt^2} = f''(t)$ *gives its* **acceleration**.

If we can measure position, then the rate of change of position is velocity, and the rate at which velocity is changing is acceleration. If the unit of position is meters, then velocity is meters/second, and acceleration is meters/second/second. Each successive derivative increases the number of "per second" qualifiers by one. This is reflected in the differential notation:

- $y = f(t)$ (no denominator)

- $\dfrac{dy}{dt} = f'(t)$ (time in denominator)

- $\dfrac{d^2y}{dt^2} = f''(t)$ (time squared in denominator)

Example 3.92   Suppose that the distance traveled by a vehicle is given by

$$s(t) = 0.02t^2 + 0.5t + 2.$$

Find the velocity $v(t)$ and acceleration $a(t)$ of the vehicle.

Solution:

Take derivatives:

$$s(t) = 0.02t^2 + 0.5t + 2 \text{ meters}$$
$$v(t) = 0.04t + 0.5 \text{ meters/second}$$
$$a(t) = 0.04 \text{ meters/second}^2$$

Velocity and acceleration are not the only rates of change we may be interested in. The next example shows how to compute the flow out of a tank from the volume of fluid in it.

Example 3.93   A cylindrical tank with an open spigot at the bottom has a volume of water that is

$$V(t) = 1216 - 72\sqrt{t} \text{ liters}$$

at a time t. Find the rate of flow of water from the tank.

Solution:

Take derivatives:

$$V(t) = 1216 - 72\sqrt{t} \text{ L}$$
$$V(t) = 1216 - 72t^{1/2}$$
$$V'(t) = -36t^{-1/2}$$
$$V'(t) = -\frac{36}{\sqrt{t}} \text{ L/sec}$$

Of course this formula only works while the tank has water in it, so we should also compute:

$$V(t) = 0$$
$$1216 - 72\sqrt{t} = 0$$
$$1216 = 72\sqrt{t}$$
$$1478656 = 5184t$$
$$t = 285.2$$

So the formula is only good for the first 285 seconds, after which the tank is empty.

3.5.1   IMPLICIT DERIVATIVES

In Section 1.4 we used the circle as an example of an interesting object that failed the vertical line test and so is not a function. On the other hand, the whole theory of derivatives we've built up works only for functions, at least so far. **Implicit derivatives** let us get around this problem. In order to use implicit derivatives we need a convenient fact about differentials: they cancel like variables.

Knowledge Box 3.19

Algebraic properties of differentials

For any variables a, b, c,

$$\frac{da}{db} \cdot \frac{db}{dc} = \frac{da}{d\!\!\!/b} \cdot \frac{d\!\!\!/b}{dc} = \frac{da}{dc}$$

and similarly

$$\frac{da/db}{dc/db} = \frac{da}{dc}$$

The trick of implicit differentiation is that we can say "y is a function of x, but we are not explicitly solving for it." Then we take the derivative of everything, as if x and y are both "the variable." Since y is a function of x, we also pick up $\dfrac{dy}{dx}$ every time we take a derivative of y. This is required by the chain rule. Let's do an example.

Example 3.94   Suppose that
$$x^2 + y^2 = 25,$$

find the tangent line at the point (3,-4).

Solution:

There is an added step to finding a tangent line this way—we need to make sure the point of tangency is legitimate (on the curve). In this case

$$(3^2) + (-4)^2 = 9 + 16 = 25$$

so we are just fine. Now we take the derivative, normally but for both variables, sticking in a $\dfrac{dy}{dx}$ each time we take a derivative of y. This yields

$$2x + 2y\frac{dy}{dx} = 0$$

Plugging in the point gives us $2(3) + 2(-4)\dfrac{dy}{dx} = 0$, and we solve to get the slope of the tangent line, $\dfrac{dy}{dx} = \dfrac{3}{4}$. Plugging into the point-slope form, $y + 4 = \dfrac{3}{4}(x - 3)$ or

$$y = \frac{3}{4}x - \frac{25}{4}.$$

Let's look at a picture of the circle and the tangent line.

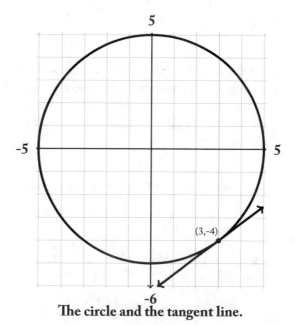

The circle and the tangent line.

Notice that we did not have to worry about whether the function we were using was the upper half of the circle or the lower half. We got a general expression that covered both halves of the circle. Which half we actually were working on was controlled by the choice of the point. (3,-4) is on the bottom half.

$$\Diamond$$

The next example permits us to use the algebraic properties of differentials.

Example 3.95   If $y = x^2$ and $x = t^2 + t + 3$, then what is $\dfrac{dy}{dt}$ when $t = 4$?

Solution:

When $t = 4$ we see that $x = 16 + 4 + 3 = 23$. Taking a couple of derivatives we see that $\dfrac{dy}{dx} = 2x$ and $\dfrac{dx}{dt} = 2t + 1$. When $t = 4$ we get that $\dfrac{dx}{dt} = 2 \cdot 4 + 1 = 9$. We can then compute that:

$$\frac{dy}{dt} = \frac{dy}{dx}\frac{dx}{dt} = 2x \cdot 9 = 2 \cdot 23 \cdot 9 = 414$$

$$\Diamond$$

One of the nice features of the implicit derivative is that it lets us avoid solving a function for y before we compute the tangent slope. The next example carefully avoids having to deal with the function

$$y = \sqrt[3]{17 - x^2}$$

Value the power!

Example 3.96   Given $x^2 + y^3 = 17$, find the tangent line to the curve at (3,2).

Solution:

First, again, check the point is on the curve. We get $3^2 + 2^3 = 9 + 8 = 17$. So, that's fine. Now compute the implicit derivative and get that:

$$2x + 3y^2\frac{dy}{dx} = 0$$

Plug in (3,2), and we get that $6 + 12\dfrac{dy}{dx} = 0$. Solving for $\dfrac{dy}{dx}$ we get the slope of the line, which is $-\dfrac{1}{2}$. Using the point slope form we get that $y - 2 = -\dfrac{1}{2}(x - 3)$, which simplifies to $y = -\dfrac{1}{2}x + \dfrac{7}{2}$. Done!

$$\Diamond$$

Suppose we have a relationship that changes with time. Then we can take the (implicit) time derivative of everything. Any variable a gets a $\dfrac{da}{dt}$ tacked on, courtesy of the chain rule.

Example 3.97   Suppose a copper disk is being heated so that its radius is expanding at 0.02 mm/min, how fast is its surface area changing when the radius is 12 mm?

Solution:

We start with the area formula for a disk:

$$A = \pi \cdot r^2$$

The variables are A and r. Taking the implicit time derivative we get:

$$\frac{dA}{dt} = 2\pi r \cdot \frac{dr}{dt}$$

The statement of the problem tells us the rate of change of the radius with respect to time is $\dfrac{dr}{dt} = 0.02$ mm/min and that $r = 12$, so we plug in and get

$$\frac{dA}{dt} = 2\pi(12)(0.02) \cong 1.51 \text{ mm}^2/\text{min}$$

$$\Diamond$$

This example is a type of problem called a **related rate** problem. In general, if we have an equation that relates quantities that vary with time, we can use its implicit time derivative to get a relationship between the way the quantities are changing.

Example 3.98   Suppose 0.5 m^3 of hydrogen is being pumped into a spherical balloon each second; how fast is the radius of the balloon changing when it holds 1000 m^3 of hydrogen?

Solution:

This problem has a couple of moving parts. First of all, the volume of a sphere, in terms of its radius, is:

$$V = \frac{4}{3}\pi r^3$$

So the (implicit) time derivative is:

$$\frac{dV}{dt} = 4\pi r^2 \frac{dr}{dt}$$

The problem asked for $\dfrac{dr}{dt}$ and gave us $\dfrac{dV}{dt} = 0.5 \text{ m}^3/\text{sec}$.

We also know we want the answer when $V = 1000$ m³, **but** our formula is in terms of the radius.

Solve:

$$1000 = \frac{4}{3}\pi r^3$$

for the radius and we get:

$$r = \sqrt[3]{\frac{3000}{4\pi}} \cong 6.2 \text{ m}$$

Now we plug in and get

$$0.5 = 4\pi (6.2)^2 \frac{dr}{dt}$$

or

$$\frac{dr}{dt} = \frac{1}{8\pi \cdot (6.2)^2} \cong 0.00104 \text{ m/sec}$$

Pretty slowly.

◊

Example 3.99   A car is approaching an intersection from the north at 100 km/hr; a second is approaching from the west at 60 km/hr. When the first car is 1 km from the intersection, and the second is 1.2 km from the intersection, how fast are the cars approaching one another?

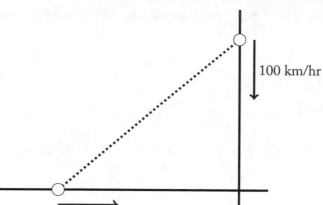

Solution:

The distance between the cars is the hypotenuse of a right triangle. If the distance of the first car from the intersection is y, and the distance of the second is x, then the distance D of the cars from one another is given by $D^2 = x^2 + y^2$.

Taking the time derivative we get:

$$2D\frac{dD}{dt} = 2x\frac{dx}{dt} + 2y\frac{dy}{dt}$$

Divide through by two and compute $D = \sqrt{1^2 + 1.2^2} = \sqrt{2.44} \cong 1.56$ km.

Plug in and get

$$2(1.56)\frac{dD}{dt} = 2(1.2)(60) + 2(1)(100)$$

or

$$\frac{dD}{dt} = \frac{344}{3.12} \cong 110.3 \text{ km/hr}$$

Notice that the answer is (and should be) plausible.

◊

3.5.2   LOGARITHMIC DIFFERENTIATION

This section highlights another use for implicit differentiation that is so cool it gets its own name: **logarithmic differentiation**. The idea is this—if a function has a lot of products and/or powers, taking the log of both sides before we take the derivative might make life easier.

Recall that:

- $\ln(ab) = \ln(a) + \ln(b)$
- $\ln\left(a^b\right) = b \cdot \ln(a)$

Example 3.100   Suppose that $y = x^5(x + 1)^7(x + 3)^3$. Find the derivative $\dfrac{dy}{dx}$.

Solution:

Take the log of both sides and take the implicit derivative.

$$y = x^5(x + 1)^7(x + 3)^3$$

$$\ln(y) = \ln(x^5(x + 1)^7(x + 3)^3)$$

$$\ln(y) = \ln(x^5) + \ln((x + 1)^7) + \ln((x + 3)^3)$$

$$\ln(y) = 5\ln(x) + 7\ln(x + 1) + 3\ln(x + 3)$$

$$\frac{1}{y} \cdot \frac{dy}{dx} = 5\frac{1}{x} + 7\frac{1}{x + 1} + 3\frac{1}{x + 3}$$

$$\frac{dy}{dx} = y\left(\frac{5}{x} + \frac{7}{x + 1} + \frac{3}{x + 3}\right)$$

$$\frac{dy}{dx} = x^5(x + 1)^7(x + 3)^3\left(\frac{5}{x} + \frac{7}{x + 1} + \frac{3}{x + 3}\right)$$

Notice that the last step eliminates the y on the right side of the equation by substituting in the known value of y in terms of x. We usually do not simplify these expressions.

Example 3.101   Given $y = \dfrac{x^6(x + 2)^4}{(x^2 + 1)^5}$, find y'.

Solution:
Take the log of both sides and take the implicit derivative.

$$y = \frac{x^6(x+2)^4}{(x^2+1)^5}$$

$$\ln(y) = 6\ln(x) + 4\ln(x+2) - 5\ln(x^2+1)$$

$$\frac{1}{y} \cdot \frac{dy}{dx} = \frac{6}{x} + \frac{4}{x+2} + 5 \cdot \frac{2x}{x^2+1}$$

$$\frac{dy}{dx} = y\left(\frac{6}{x} + \frac{4}{x+2} + \frac{10x}{x^2+1}\right)$$

$$\frac{dy}{dx} = \frac{x^6(x+2)^4}{(x^2+1)^5}\left(\frac{6}{x} + \frac{4}{x+2} + \frac{10x}{x^2+1}\right)$$

$\Diamond$

Logarithmic differentiation also permits us to take the derivative of functions that we couldn't otherwise work with at all. The next example demonstrates this.

Example 3.102   Find the derivative of $y = x^x$.

Solution:

$$y = x^x$$

$$\ln(y) = x \cdot \ln(x)$$

$$\frac{1}{y} \cdot \frac{dy}{dx} = x \cdot \frac{1}{x} + 1 \cdot \ln(x)$$

$$\frac{dy}{dx} = y\,(1 + \ln(x))$$

$$\frac{dy}{dx} = x^x\,(1 + \ln(x))$$

$\Diamond$

Of course, once we have the ability to take derivatives, we can find tangent lines, do related relate problems, or anything else one can do with derivatives.

PROBLEMS

Problem 3.103   Each of the functions below gives distance as a function of time. Compute velocity and acceleration.

1. $f(t) = 0.02t^3 + t^2$

2. $g(t) = t \cdot \ln(t)$

3. $h(t) = \dfrac{1}{t^2 + 1}$

4. $r(t) = \sin(t)$

5. $s(t) = \tan(t)$

6. $q(t) = \left(\sqrt{t} + 1\right)^{3.6}$

Problem 3.104   The height of a ball above the ground is given by the equation

$$h(t) = 112 + 6t - 5t^2 \ m$$

1. How high in the air is the ball at time $t = 0$?

2. Is the ball moving up or down at $t = 0$?

3. When does the ball hit the ground?

4. Find an expression for the ball's velocity.

5. What is the ball's acceleration?

Problem 3.105   The amount of water in a tank at time t, in liters, is given by

$$A(t) = 0.2t + \sqrt{t}$$

1. At what rate is water flowing into the tank?

2. Is the tank filling or emptying?

3. At what point will there be 100 l of water in the tank?

Problem 3.106   Given a falling object that experiences constant, non-zero acceleration, explain why its position function is a quadratic function.

Problem 3.107   Does a ball falling from a great height above the earth experience constant acceleration?

Problem 3.108   Given the pairs of functions, find the indicated differential.

1. $y = 3x + 5$, $x = t^2 - 1$

   Find $\dfrac{dy}{dt}$ at $t = 2$.

2. $y = \ln(x)$, $x = 5t + 4$

   Find $\dfrac{dy}{dt}$ at $t = 1$.

3. $y = \dfrac{x+1}{x-1}$, $x = \sin(t)$

   Find $\dfrac{dy}{dt}$ at $t = \dfrac{\pi}{3}$.

4. $y = x(x + 1)$, $x = t^3 + t^2 + 2t + 1$

   Find $\dfrac{dy}{dt}$ at $t = -2$.

5. $y = e^x$, $x = t^2 + 51$

   Find $\dfrac{dy}{dt}$ at $t = 1$.

6. $y = 4x - 1$, $x = t \cdot e^{-t}$

   Find $\dfrac{dy}{dt}$ at $t = 1$.

Problem 3.109   Suppose $\dfrac{dy}{dx} = 1.2$, $\dfrac{dx}{dt} = 3.4$, and $\dfrac{dt}{du} = -1.5$. Find the following differentials.

1. $\dfrac{dy}{dt}$ 2. $\dfrac{dy}{du}$ 3. $\dfrac{dx}{du}$ 4. $\dfrac{dt}{dy}$ 5. $\dfrac{du}{dy}$ 6. $\dfrac{du}{dx}$

Problem 3.110   Find the tangent line to each of the following curves at the specified point.

1. $x^2 + y^2 = 20$ at $(-4, 2)$

2. $(xy + 1)^2 = 9$ at $(2, -2)$

3. $x^2 + 3xy + y^2 = 5$ at $(1, 1)$

4. $x \cdot \cos(y) = 1$ at $\left(\sqrt{2}, \dfrac{\pi}{4}\right)$

5. $x^5 y + xy^5 + 1 = 3$ at $(-1, -1)$

6. $x^3 + y^3 + 2xy + 4 = 17$ at $(1, 2)$

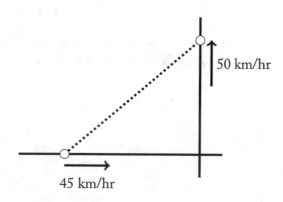

50 km/hr

45 km/hr

Problem 3.111   A car is leaving an intersection northward at 50 km/hr; another is coming from the west at 45 km/hr. When the northward car is 100 m from the intersection, and the eastward car is 60 m from the intersection, what is the relative velocity of the cars?

Problem 3.112   A small metal disk is cooled with cold water so that its diameter is decreasing by 0.003 mm/sec. How fast is the surface area of one side decreasing when the diameter is 20 mm?

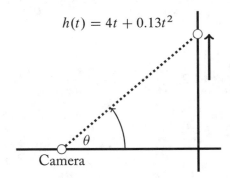

Problem 3.113   Suppose that the height of an object at time t is given by $h(t) = 4t + 0.13t^2$ m. The object is being tracked by a camera 10 m away. Find an expression for the rate of rotation of the camera at time t.

Problem 3.114   The top of a 5 m ladder is sliding down a wall at 1.4 m/sec. How fast is the foot of the ladder sliding away from the wall when the top of the ladder is 4 m off the floor?

Problem 3.115   If three cubic meters of air per second are pumped into a spherical balloon, find the rate at which its diameter and surface area are changing when it holds 400 m^3 of air. Recall that for a sphere,

$$V = \frac{4}{3}\pi r^3$$

and

$$A = 4\pi r^2$$

Problem 3.116   Find $\dfrac{dy}{dx}$ for each of the following functions.

1. $y = (x^2 + 1)^4 (x^2 + x + 1)^5$

2. $y = \dfrac{x^6}{(x + 1)^{12}(x + 4)^5}$

3. $y = e^x (x^2 + 1)^7 (x^2 - 2)^5$

4. $y = \left(x^2 + 1\right)^x$

5. $y = x^{\tan(x)}$

6. $y = (\sin(x) + 2)^x$

7. $y = x^8 \cdot \tan^5(x) \cdot \sec^7(x)$

8. $y = x(x + 1)(x + 2)(x + 3)(x + 4)(x + 5)$

Problem 3.117   Find the tangent line to

$$y = \frac{x^6(x^2 + 1)}{(x^2 + x + 1)^5}$$

at the point (-1,2).

Problem 3.118   Find the tangent line to

$$y = x^{x^2+1}$$

at the point (2,32).

Problem 3.119   Find the tangent line to

$$y = (x + 1)^{x+3}$$

at the point (1,16).

Problem 3.120   If two cars are approaching a right-angle intersection at the same speed S and are the same distance away from it, how fast are they approaching one another?

CHAPTER 4

Curve Sketching

This chapter explores using the derivative to sketch curves. A sketch is not exact in it measurements, but it captures the sense of the shape of a curve and, if the function that generates the curve is one that expresses a physical law, helps us to understand the physics. We can always graph a function with a computer or calculator, but this will give us an exact graph that may hide important features that are smaller scale. The discipline of curve sketching grants a second point of view on the shape and character of a curve.

4.1 LIMITS AT INFINITY

We start curve sketching by using roots (places where a curve crosses the x-axis) and **vertical** and **horizontal** asymptotes. Finding horizontal asymptotes will require us to compute limits of a function as x grows infinitely far from zero—in either of the possible directions. The first section of the chapter is about **limits at infinity**.

The limit at positive infinity is the value a function gets close to as x gets larger and larger. The limit at negative infinity is the value a function gets close to as x gets smaller and smaller. The absolute value function gives us the ability to figure out if two things are "close." If $|f(x) - L|$ is small, then $f(x)$ is close to L. Traditionally, the Greek letter epsilon, ϵ, is used to represent small values. The situation is illustrated in Figure 4.1.

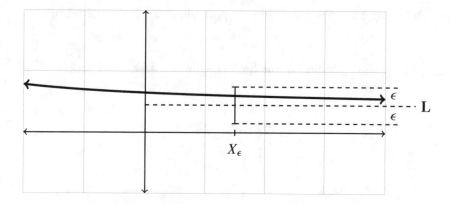

Figure 4.1: A limit at infinity: as long as $x > X_\epsilon$ we have that $f(x)$ is closer to L than ϵ.

This gives us the following formal definitions.

Definition 4.1 *If we can find a number L so that for every number $\epsilon > 0$ we can find a value X_ϵ so that when $x \geq X_\epsilon$ we have that $|f(x) - L| < \epsilon$, then L is the limit of $f(x)$ as $x \to \infty$.*

Definition 4.2 *If we can find a number L so that for every number $\epsilon > 0$ we can find a value X_ϵ so that when $x \leq X_\epsilon$ we have that $|f(x) - L| < \epsilon$, then L is the limit of $f(x)$ as $x \to -\infty$.*

There are a number of cases where we can compute the limit using algebra. It is entirely possible that no such number as L exists. Examine the sine function. As $x \to \infty$ it continues to alternate back and forth, settling down to no value L.

A simpler way for the limit to fail to exist is for it to **diverge**. This means that as x gets bigger and bigger $f(x)$ does as well. Or, more formally:

Definition 4.3 *If for any constant c, we can find X_c so that $x > X_c$ means $f(x) \geq c$, then*

$$\lim_{x \to \infty} f(x)$$

diverges to infinity.

We will begin by doing examples where we compute limits using the definition.

Example 4.1   Using the definition of the limit (Definition 4.1), show that:

$$\lim_{x \to \infty} \frac{1}{1 + x^2} = 0$$

Solution:

For each $\epsilon > 0$ we need to find X_ϵ. Assume we have fixed ϵ and compute X_ϵ. Remember that when we take the reciprocal of both sides of an inequality, it reverses.

$$\left| \frac{1}{1 + X_\epsilon^2} - 0 \right| < \epsilon$$

$$\frac{1}{1 + X_\epsilon^2} < \epsilon$$

$$1 + X_\epsilon^2 > \frac{1}{\epsilon}$$

$$X_\epsilon^2 > \frac{1}{\epsilon} - 1$$

$$X_\epsilon > \sqrt{\frac{1}{\epsilon} - 1}$$

So the value $X_\epsilon = \sqrt{\frac{1}{\epsilon} - 1}$ meets the requirements of the definition for $\epsilon \leq 1$. If $\epsilon > 1$, this value is undefined because it involves the square root of a negative number, but, since the function has a maximum value of 1 (at zero), satisfying the inequality for larger values of ϵ is easy.

◊

Example 4.2   Show, with the definition of divergence (Definition 4.3) that $\lim_{x \to \infty} x^3$ diverges to ∞.

Solution:
For a constant c, set $X_c = \sqrt[3]{c}$. If

$$x \geq X_c$$

$$x \geq \sqrt[3]{c}$$

$$x^3 \geq c$$

$$f(x) \geq c$$

and we have the desired property for the given X_c.

◊

Now let's solve the limit from Example 4.1 algebraically. To do this we need to accept the fact that a constant quantity divided by a quantity that grows without limit is, itself, going to zero. In this case:

Knowledge Box 4.1

Limit of a constant over a growing power of x

$$\lim_{x \to \infty} \frac{1}{x^\alpha} = 0 \; for \; \alpha > 0$$

Example 4.3   Compute

$$\lim_{x \to \infty} \frac{1}{x^2 + 1}$$

using the fact in Knowledge Box 4.1.

Solution:

$$\lim_{x \to \infty} \frac{1}{x^2 + 1} = \lim_{x \to \infty} \left(\frac{\frac{1}{x^2}}{\frac{1}{x^2}} \right) \frac{1}{x^2 + 1}$$

$$= \lim_{x \to \infty} \frac{\frac{1}{x^2}}{\frac{x^2}{x^2} + \frac{1}{x^2}}$$

$$= \lim_{x \to \infty} \frac{\frac{1}{x^2}}{1 + \frac{1}{x^2}}$$

$$= \frac{0}{1 + 0}$$

$$= 0$$

When all the x's vanish we are left with a constant that does not depend on x. The limit of a constant is the constant itself.

◊

More generally, this trick (divide top and bottom by the highest power of x) works whenever we have a polynomial divided by a polynomial. Let's do another example.

Example 4.4   Find:

$$\lim_{x \to \infty} \frac{2x^3}{x^3 + 3x + 1}$$

Solution:

$$\lim_{x \to \infty} \frac{2x^3}{x^3 + 3x + 1} = \lim_{x \to \infty} \frac{1/x^3}{1/x^3} \cdot \frac{2x^3}{x^3 + 3x + 1}$$

$$= \lim_{x \to \infty} \frac{2}{1 + \dfrac{3}{x^2} + \dfrac{1}{x^3}}$$

$$= \frac{2}{1 + 0 + 0}$$

$$= 2$$

$\Diamond$

This leads to a simple rule for the ratio of two polynomials.

Knowledge Box 4.2

Limit of the ratio of polynomials

Suppose that $f(x)$ is a polynomial of degree n and $g(x)$ is a polynomial of degree m. Then

$$\lim_{x \to \infty} \frac{f(x)}{g(x)} \text{ is:}$$

- *0 if $m > n$*

- *∞ if $n > m$*

- *Equal to the ratio of the coefficients of the highest degree terms on the top and bottom if $n = m$*

Example 4.5

$$\lim_{x\to\infty} \frac{3x^2 + 5x + 1}{x^2 + 2x + 7} = 3$$

Notice how the rule in Knowledge Box 4.2 makes this sort of problem have a one-step immediate solution?

This technique for resolving limits at infinity can be used more generally by making the rule about a constant over something growing more general.

Knowledge Box 4.3

Limit of a constant over a growing quantity

Suppose that $\lim\limits_{x\to\infty} f(x) = \infty$ *and that C is a constant. Then*

$$\lim_{x\to\infty} \frac{C}{f(x)} = 0.$$

With this new fact, the rule changes from "divide by the highest power of x" to "divide by the fastest growing quantity."

Example 4.6   Compute:

$$\lim_{x\to\infty} \frac{e^x}{1 + 2e^x}$$

Solution:

$$\lim_{x\to\infty} \frac{e^x}{1 + 2e^x} = \lim_{x\to\infty} \frac{e^{-x}}{e^{-x}} \cdot \frac{e^x}{1 + 2e^x}$$

$$= \lim_{x \to \infty} \ \frac{e^{-x} \cdot e^x}{e^{-x} + e^{-x} \cdot 2e^x}$$

$$= \lim_{x \to \infty} \ \frac{1}{e^{-x} + 2}$$

$$= \lim_{x \to \infty} \ \frac{1}{\dfrac{1}{e^x} + 2}$$

$$= \frac{1}{0 + 2}$$

$$= \frac{1}{2}$$

$$\Diamond$$

Since there is only one fast-growing quantity in this example (e^x) – it's not hard to spot the fastest growing quantity. The next Knowledge Box has information about which types of functions grow faster. Except for "constants" all the functions in Knowledge Box 4.4 grow to infinity as x does.

Knowledge Box 4.4

Which functions grow faster?

- *Logarithms grow faster than constants*

- *Positive powers of x grow faster than logarithms*

- *Larger positive powers of x grow faster than smaller positive powers of x*

- *Exponentials (with positive exponents) grow faster than positive powers of x*

- *Exponentials with larger exponents grow faster than those with smaller exponents*

Let's do some one-step examples with these new rules.

Example 4.7

$$\lim_{x \to \infty} \frac{3}{\ln(x) + 1} = 0$$

$$\lim_{x \to \infty} \frac{x^5}{e^x + 1} = 0$$

$$\lim_{x \to \infty} \frac{e^{2x}}{1 + e^x} = \infty$$

$$\Diamond$$

Aside from keeping track of signs, limits as

$$x \to -\infty$$

obey the same rules as those going to positive infinity. With limits at infinity under some degree of control, we can now discuss asymptotes.

Definition 4.4 *An **asymptote** is a horizontal or vertical line that the graph of a function approaches arbitrarily close to but does not touch.*

The function shown in Figure 4.2 has a horizontal asymptote (dashed line) and a vertical one, dotted line, as well as a **root** at (0,0).

Knowledge Box 4.5

Rules for finding asymptotes

The **horizontal asymptotes** *to the graph of* $f(x)$ *are the y values:*

$$y = \lim_{x \to \pm\infty} f(x)$$

There are at most two horizontal asymptotes. **Vertical asymptotes** *occur whenever there is a point* $x = v$ *where* $f(v)$ *fails to exist because of a division of a non-zero quantity by zero.*

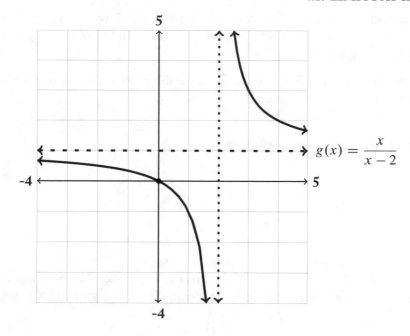

Figure 4.2: A function with a vertical and horizontal asymptote.

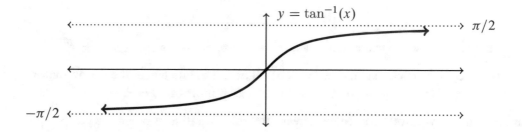

Figure 4.3: The inverse tangent function, two horizontal asymptotes.

The example function in Figure 4.2 has a horizontal asymptote at $y = 1$, because

$$\lim_{x \to \pm\infty} \frac{x}{x - 2} = 1$$

and a vertical asymptote at $x - 2 = 0$ or $x = 2$ where a divide-by-zero occurs. It also has a root at $x = 0$.

Figure 4.3 shows an example of a function with two horizontal asymptotes: $g(x) = \tan^{-1}(x)$. The function approaches $\pi/2$ as x goes to positive infinity and $-\pi/2$ as x goes to $-\infty$. The function also has a single root at (0,0).

When sketching the graph of a function the first three things we list are the roots, the horizontal asymptotes, and the vertical asymptotes.

Example 4.8   Find the roots and asymptotes of

$$f(x) = \frac{3 - x^2}{x^2 - 4}.$$

Solution:

First of all, we need to find the roots by solving for those x-values that yield a y value of zero. An important part of this is to remember that a fraction is zero only when its numerator is zero at places where its denominator is not zero.

$$3 - x^2 = 0$$
$$3 = x^2$$
$$x = \pm\sqrt{3}$$

Since neither of these are values that make $x^2 - 4$ zero, our roots are at $x = \pm\sqrt{3}$.

Next we need the vertical asymptotes, which will be the points where the denominator is zero; $x^2 - 4 = (x + 2)(x - 2) = 0$, so there are vertical asymptotes at $x = \pm 2$.

The horizontal asymptotes can be found with the rule for ratios of polynomials:

$$\lim_{x \to \infty} \frac{3 - x^2}{x^2 - 4} = \frac{-1}{1} = -1$$

So we have a horizontal asymptote at $y = -1$. For now, plot a few points near a vertical asymptote to figure out which way the curve is going as you approach the asymptote from either side. The next section will give us another tool to figure this out.

Now, we can sketch the graph of $f(x)$. Notice that roots and both sorts of asymptotes are shown.

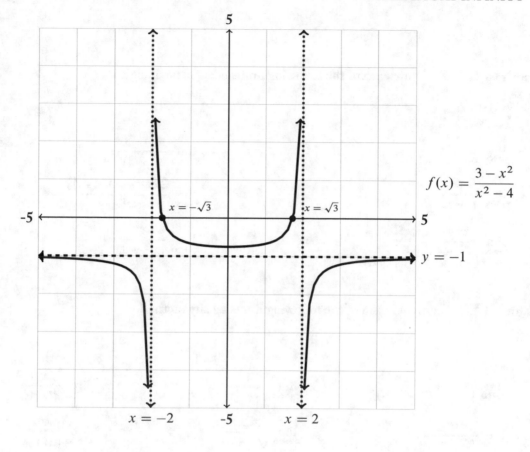

$$f(x) = \frac{3 - x^2}{x^2 - 4}$$

$\Diamond$

PROBLEMS

Problem 4.9   Using the definition of a limit at infinity show that

$$\lim_{x \to \infty} \frac{2}{x^2 + 4} = 0.$$

Problem 4.10   Using the definition of a limit at infinity show that

$$\lim_{x \to \infty} \frac{x + 1}{x - 1} = 1.$$

Problem 4.11   Compute each of the following limits using explicit algebra.

1. $\lim\limits_{x\to\infty} \dfrac{x}{2x+1}$

2. $\lim\limits_{x\to\infty} \dfrac{x^2}{1+5x}$

3. $\lim\limits_{x\to\infty} \dfrac{2x^2+3x+5}{x^2-4x+6}$

4. $\lim\limits_{x\to\infty} \dfrac{x^3+3x+2}{x^3+3x^2+2}$

5. $\lim\limits_{x\to\infty} \dfrac{3x(x+1)(x-1)}{x^3+4}$

6. $\lim\limits_{x\to\infty} \dfrac{5x^2}{4x^2+1}$

Problem 4.12   Compute each of the following limits by any method.

1. $\lim\limits_{x\to\infty} \dfrac{3e^x}{1+e^x}$

2. $\lim\limits_{x\to\infty} \dfrac{5e^x}{1+e^{2x}}$

3. $\lim\limits_{x\to\infty} \dfrac{x^2}{x\cdot\ln(x)}$

4. $\lim\limits_{x\to\infty} \dfrac{\ln(x)}{\sqrt{x}+1}$

5. $\lim\limits_{x\to\infty} \dfrac{\sqrt{x}+2}{\sqrt[3]{x}+1}$

6. $\lim\limits_{x\to\infty} \dfrac{e^x}{\ln(x^5)}$

Problem 4.13   Find the root and asymptotes (vertical and horizontal) for each of the following functions.

1. $f(x) = \dfrac{x^2-1}{x}$

2. $g(x) = \dfrac{x}{x^2-1}$

3. $h(x) = \ln(1/x)$

4. $r(x) = \dfrac{e^x-1}{e^x+1}$

5. $s(x) = \dfrac{2x^2-8}{x^2-9}$

6. $q(x) = \dfrac{1}{x} + \dfrac{1}{x-2}$

7. $a(x) = \tan^{-1}(x^2)$

8. $b(x) = \dfrac{x^3+6x^2+11x+6}{x^3-6x^2+11x-6}$

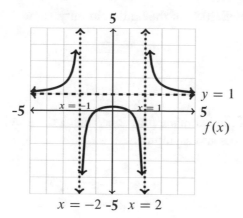

Problem 4.14   For the graph above, find the roots and the vertical and horizontal asymptotes.

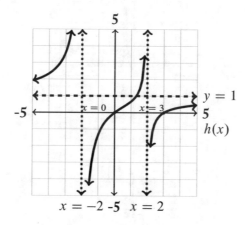

Problem 4.15   For the graph shown above, assume that the function plotted is the ratio of two quadratic equations.

1. List the roots

2. List the vertical asymptotes

3. List the horizontal asymptotes

4. Give a sensible formula for $h(x)$

Problem 4.16   Using the definition of the limit at infinity, show $\lim_{x \to \infty} x^2$ diverges to ∞.

Problem 4.17   Using the definition of the limit at infinity, show

$$\lim_{x \to \infty} \frac{x^2 - 1}{x}$$

diverges to ∞.

Problem 4.18   For each of the following functions, sketch the graph to the best of your ability, showing and labeling the roots and asymptotes.

1. $\lim\limits_{x \to \infty} \dfrac{2x}{x + 1}$

2. $\lim\limits_{x \to \infty} \dfrac{x^2}{1 + 5x^2}$

3. $\lim\limits_{x \to \infty} \dfrac{x^2 - 9}{x^2 - 1}$

4. $\lim\limits_{x \to \infty} \dfrac{x^2 + 3x}{x^2 - 1}$

5. $\lim\limits_{x \to \infty} \dfrac{x^4}{x^4 - 1}$

6. $\lim\limits_{x \to \infty} \dfrac{5x^2}{x^2 - 4}$

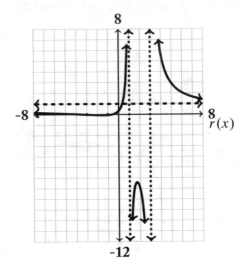

Problem 4.19   For the graph shown above, assume that the function plotted is the ratio of two quadratic equations.

1. List the roots

2. List the vertical asymptotes

3. List the horizontal asymptotes

4. Give a sensible formula for $r(x)$

Problem 4.20   Describe the roots and asymptotes of

$$f(x) = \frac{1}{x^n}$$

where n is a positive whole number including the way the function approaches the asymptotes. Hint: there are two outcomes.

Problem 4.21   If $f(x)$ and $g(x)$ are both polynomials, what is the largest number of horizontal asymptotes that

$$h(x) = \frac{f(x)}{g(x)}$$

can have? Explain your answer carefully.

Problem 4.22   Find the number of horizontal asymptotes of

$$f(x) = \frac{|x^3|}{x^3 + 1}$$

4.2   INFORMATION FROM THE DERIVATIVE

There are two sorts of useful information for sketching a curve that we can pull out of the derivatives of a function. We can compute where it is **increasing** and **decreasing,** and we can compute where it is curved up (**concave up**) or curved down (**concave down**).

4.2.1   INCREASING AND DECREASING RANGES

Remember that a derivative is a rate of change. This means that when $f'(x) > 0$ in a range, the function is increasing in that range, and when $f'(x) < 0$ the function is decreasing in that range. Let's nail down exactly what it means to be increasing or decreasing on a range.

Definition 4.5 *A function is* **increasing** *on an interval if, for each $u < v$ in the interval $f(u) < f(v)$.*

Definition 4.6 *A function is* **decreasing** *on an interval if, for each $u < v$ in the interval $f(u) > f(v)$.*

Now we are ready for the derivative-based rules on when a function is increasing or decreasing.

Knowledge Box 4.6

Derivative-based function rules

- *A function $f(x)$ is* **increasing** *where $f'(x) > 0$*
- *A function $f(x)$ is* **decreasing** *where $f'(x) < 0$*
- *Those $x = c$ where $f'(c) = 0$ are called* **critical values**
- *The points $(c, f(c))$ where $f'(c) = 0$ are called* **critical points**

Example 4.23   Find the critical point(s) and increasing and decreasing ranges for

$$f(x) = x^2 - 4.$$

Solution:

We see $f'(x) = 2x$. Solving $2x = 0$ we get that $x = 0$ is the only critical value. So the critical point is $(0, -4)$.

Solving $2x < 0$ we see $f'(x) < 0$ on $(-\infty, 0)$; similarly $f'(x) > 0$ on $(0, \infty)$.

The following graph permits us to check all this against the actual behavior of the function.

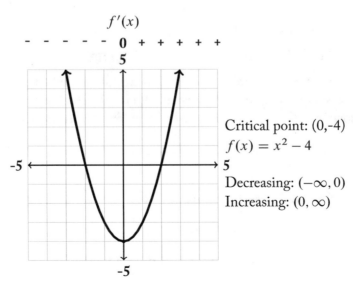

Critical point: (0,-4)
$f(x) = x^2 - 4$

Decreasing: $(-\infty, 0)$
Increasing: $(0, \infty)$

Notice that we have a row of +, -, 0 symbols across the top of the graph: these show the sign of $f'(x)$ and are handy for analysis of increasing and decreasing ranges, as we will see in a minute.

$$\Diamond$$

Example 4.24   Find the critical point(s) and increasing and decreasing ranges for

$$f(x) = x^3 - 4x.$$

Solution:

Finding the increasing and decreasing ranges requires that we first find the critical values, where $f'(c) = 0$.

$$f'(x) = 3x^2 - 4$$

$$3x^2 - 4 = 0$$

$$3x^2 = 4$$

$$x^2 = \frac{4}{3}$$

$$x = \pm\frac{2}{\sqrt{3}}$$

The derivative can only change between positive and negative at $c = \pm\frac{2}{\sqrt{3}}$, so we plug in values in each of the resulting ranges. The value $c = \frac{2}{\sqrt{3}}$ is a little larger than one, so let's look at -2 and 2. We can make a table of values.

x	$f'(x)$	$\pm$
$-\infty$	na	na
-2	$8 > 0$	+
$-\dfrac{2}{\sqrt{3}}$	0	0
0	$-4 < 0$	-
$\dfrac{2}{\sqrt{3}}$	0	0
2	$8 > 0$	+
∞	na	na

So, the function is:

Increasing on: $(-\infty, -\frac{2}{\sqrt{3}}) \cup (\frac{2}{\sqrt{3}}, \infty)$

Decreasing on: $(-\frac{2}{\sqrt{3}}, \frac{2}{\sqrt{3}})$

with critical points at $\left(\pm\frac{2}{\sqrt{3}}, \mp\frac{16}{3\sqrt{3}}\right)$

Notice that the use of $\mp$ means that the sign of the second coordinate of the critical points is the opposite of the sign of the first. We also usually use a much more compact form for the table of signs given above:

$$f'(x): \;\; (-\infty) + + + \left(-\frac{2}{\sqrt{3}}\right) - - - \left(\frac{2}{\sqrt{3}}\right) + + + (\infty)$$

Numbers inserted into the chain of "+" and "−" symbols represent critical points. This device is called a **sign chart** for increasing and decreasing ranges. Below is a picture of the function with the features we just located shown. The critical points are plotted.

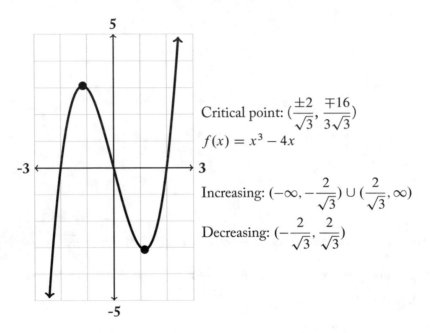

Critical point: $(\frac{\pm 2}{\sqrt{3}}, \frac{\mp 16}{3\sqrt{3}})$

$f(x) = x^3 - 4x$

Increasing: $(-\infty, -\frac{2}{\sqrt{3}}) \cup (\frac{2}{\sqrt{3}}, \infty)$

Decreasing: $(-\frac{2}{\sqrt{3}}, \frac{2}{\sqrt{3}})$

$\Diamond$

An alert reader will have noticed that we carefully avoided, so far in this section, examples involving asymptotes. The reason for this is that they can influence the increasing/decreasing ranges as well.

Knowledge Box 4.7

A continuous, differentiable function can only change between increasing and decreasing at a critical value or at a vertical asymptote.

This means that we include the position of vertical asymptotes along with critical values on the sign chart for finding increasing and decreasing ranges for a function.

Example 4.25 Find the critical points, vertical asymptotes, and increasing and decreasing ranges for

$$f(x) = \frac{x^2 - 1}{x^2 - 4}$$

Solution:
The vertical asymptotes are easy: $x^2 - 4 = 0$ at $x = \pm 2$. The critical points require us to solve $f'(x) = 0$, which gives us:

$$\frac{(x^2 - 1)(2x) - (x^2 - 4)(2x)}{(x^2 - 4)^2} = 0$$
$$(x^2 - 1)(2x) - (x^2 - 4)(2x) = 0$$
$$2x^3 - 2x - 2x^3 + 8x = 0$$
$$6x = 0$$
$$x = 0$$

So there is a critical value at $x = 0$ and a critical point at $(0, \frac{1}{4})$. Remember that a fraction is zero only where its numerator is zero. Let's make the sign chart with the critical values and vertical asymptotes. We can plug in any value in an interval to get the $\pm$ value for $f'(x)$:

$$(-\infty) + + + (-2) + + + (0) - - - (2) - - - (\infty)$$

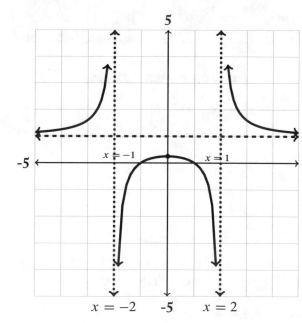

$$f(x) = \frac{x^2 - 1}{x^2 - 4}$$
Critical point: $(0, 1/4)$

Increasing: $(-\infty, -2) \cup (-2, 0)$

Decreasing: $(0, 2) \cup (2, \infty)$

Notice that the function can increase both up to and after the asymptote; but at the asymptote the function goes to $+\infty$ or comes from $-\infty$. It does not increase through the asymptote. This means that these points are **gaps** in the sign chart and the increasing/decreasing ranges.

4.2.2   CONCAVITY

This section introduces the last quantity we will be using for sketching the graphs of functions: concavity. Concavity—as we learned in Chapter 3—depends on the sign of the second derivative. We will use sign charts for concavity, just as for increasing and decreasing values.

Knowledge Box 4.8

Concavity rules

- *Where $f''(x) > 0$ we say a function is* **concave up**
- *Where $f''(x) < 0$ we say a function is* **concave down**
- *Values $x = c$ where $f''(c) = 0$ are* **inflection values**
- *Points $(c, f(c))$ there $f''(c) = 0$ are* **inflection points**

Example 4.26   Find the inflection points and concave up and down ranges of

$$f(x) = x^3 - 4x.$$

Solution:

We need the second derivative:

$$f'(x) = 3x^2 - 4$$
$$f''(x) = 6x$$

Solve $6x = 0$, and we get that there is an inflection value at $x = 0$ and an inflection point at $(0,0)$. Let's make a sign chart plugging in $f''(-1) = -6$ and $f''(1) = 6$ to get:

$$(-\infty) - - - (0) + + + (\infty)$$

Another way to understand where a curve is concave up or down is that the shape of the curve holds water where it is concave up and does not where it is concave down. Verify the information about concavity and inflection points on the following graph.

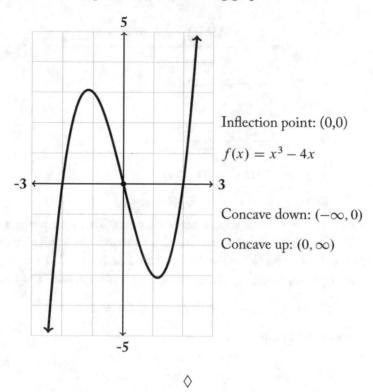

Inflection point: (0,0)

$f(x) = x^3 - 4x$

Concave down: $(-\infty, 0)$

Concave up: $(0, \infty)$

◊

Concavity can also change at vertical asymptotes. So they are included on the second derivative sign chart for concavity. Let's do an example that shows this phenomenon.

Example 4.27   Find the inflection points, concave up, and concave down ranges for

$$f(x) = \frac{x^2 - 1}{x^2 - 4}.$$

Solution:

We've already computed, in Example 4.25, that

$$f'(x) = \frac{6x}{(x^2 - 4)^2}.$$

For inflection and concavity we need $f''(x)$.

$$f''(x) = \frac{\left(x^2 - 4\right)^2 (6) - 6x \cdot 2\left(x^4 - 4\right)^1 (2x)}{\left(x^2 - 4\right)^4}$$

$$= \frac{\left(x^2 - 4\right)^{\cancel{2}1} (6) - 6x \cdot 2\cancel{\left(x^4 - 4\right)}(2x)}{\left(x^2 - 4\right)^{\cancel{4}3}}$$

$$= \frac{6\left(x^2 - 4\right) - 24x^2}{\left(x^2 - 4\right)^3}$$

$$= \frac{6x^2 - 24 - 24x^2}{\left(x^2 - 4\right)^3}$$

$$= -\frac{18x^2 + 24}{\left(x^2 - 4\right)^3}$$

Which is zero when $18x^2 + 24 = 0$, which never happens. This means the function has no inflection points. We have already computed that this function has vertical asymptotes at $x = \pm 2$. The sign chart for the second derivative is thus based on $x = \pm 2$. Plugging in $x = \pm 3$ to $f''(x)$ we get a positive result; plugging in zero we get a negative result. The sign chart is thus:

$$f''(x) : (-\infty) + + + (-2) - - - (2) + + + (\infty)$$

Check this result against the graph.

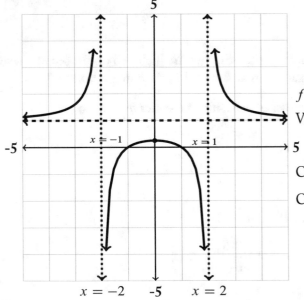

$$f(x) = \frac{x^2 - 1}{x^2 - 4}$$

Vertical asymptotes: $x = \pm 2$

Concave up: $(-\infty, -2) \cup (2, \infty)$

Concave down: $(-2, 2)$

It's worth noting that the "holds water" analogy for concave up breaks down on this example.

$$\Diamond$$

PROBLEMS

Problem 4.28   For each of the following functions, find the critical point(s).

1. $f(x) = \dfrac{x^2 - 4}{x}$

2. $g(x) = \dfrac{1}{x^2 - 4x + 5}$

3. $h(x) = \sin(x)$

4. $r(x) = x^2 + \dfrac{1}{x}$

5. $s(x) = \dfrac{x^3 - 1}{x}$

6. $q(x) = \sec(x)$

Problem 4.29   For each of the following functions, find the critical points and the vertical asymptotes, if any. Based on these, make a sign chart for the first derivative and report the ranges on which the function is increasing and decreasing.

1. $f(x) = x^2 + 4x + 5$

2. $g(x) = \dfrac{x^2 - 1}{x^2}$

3. $h(x) = \dfrac{x}{x^2 - x - 6}$

4. $r(x) = \ln(x^2 + 3)$

5. $s(x) = \sin(x)$

6. $q(x) = \dfrac{1}{x} - \dfrac{1}{x^2}$

Problem 4.30   Find a cubic polynomial that is never decreasing.

Problem 4.31   Find a cubic polynomal function that has critical values at $x = \pm 5$.

Problem 4.32   Find a cubic polynomal function that has critical values at $x = -2$ and $x = 4$.

Problem 4.33   Show that a polynomial of degree n has *at most* $n - 1$ critical points.

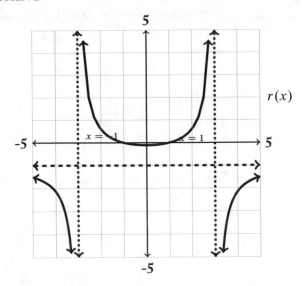

Problem 4.34   For the graph above, make the best assessment you can of the critical points, vertical asymptotes, increasing ranges, and decreasing ranges.

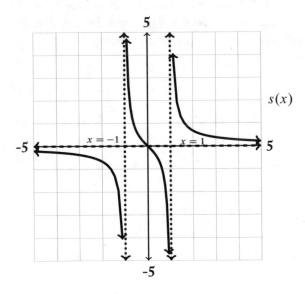

Problem 4.35   For the graph above, make the best assessment you can of the critical points, vertical asymptotes, increasing ranges, and decreasing ranges.

Problem 4.36   For each of the following functions, find the inflection point(s).

1. $f(x) = \dfrac{1}{x^2 + 1}$

2. $g(x) = x^3 - 3x^2 + 7x + 152$

3. $h(x) = \cos(x)$

4. $r(x) = \ln(x^2 + 1)$

5. $s(x) = \dfrac{x^2 - 1}{x^2 + 1}$

6. $q(x) = \tan^{-1}(x)$

Problem 4.37   For each of the following functions, find the inflection point(s) and vertical asymptotes, make a sign chart for concavity, and give the concave up and down ranges.

1. $f(x) = x^4 - 3x^2 + 2$

2. $g(x) = x^3 - 6x^2 + 5x + 6$

3. $h(x) = x^4 - x^2 + 6$

4. $r(x) = \ln(x^2 + 4)$

5. $s(x) = \dfrac{x^2 - 4}{x^2 + 4}$

6. $q(x) = \tan^{-1}(x^2)$

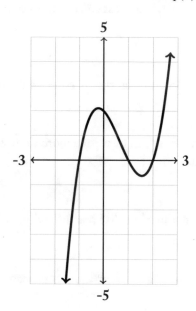

Problem 4.38   For the graph above, do the best you can to find the inflection points and concave up and concave down intervals.

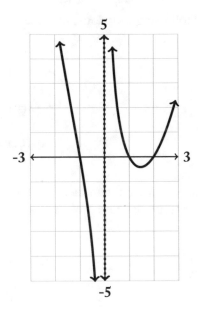

Problem 4.39   For the graph above, do the best you can to find the inflection points and concave up and concave down intervals.

Problem 4.40   Find a quadratic polynomial that is concave down everywhere.

Problem 4.41   Find a polynomal function of degree more than two that has only one concavity value – excluding inflection points.

Problem 4.42   Find a polynomal function that has inflection values at $x = -4$ and $x = 2$.

Problem 4.43   Show that a polynomial of degree n has *at most* $n - 2$ inflection points.

Problem 4.44   Suppose that $f(x)$ is a polynomial function with ranges that are both concave up and concave down. What is the smallest degree $f(x)$ can have?

4.3   THE FULL REPORT FOR CURVE SKETCHING

Knowledge Box 4.9

The table for a full report

Roots:

Vertical asymptotes:

Horizontal asymptotes:

Critical points:

Increasing on:

Decreasing on:

Inflection points:

Concave up on:

Concave down on:

In this chapter so far we have found nine types of information that help us sketch the curve generated by a function. These are the roots, both sorts of asymptotes, the critical points together with the increasing and decreasing ranges, and the inflection points together with the concave up and concave down ranges. In this section all we are going to do is pull them together into a single type of report. In addition to a good, annotated sketch of a curve, the information in Knowledge Box 4.9 is used to make a full report on a function.

With all the pieces in place, let's do a full example on a fairly simple curve. This first example will skip the issue of asymptotes by not having any.

Example 4.45   Make a full report and sketch of the curve

$$y = \frac{1}{3}x^3 - 3x.$$

Solution:

Factor to find roots:

$$\frac{1}{3}x^3 - 3x = \frac{1}{3}x(x^2 - 9) = \frac{1}{3}x(x - 3)(x + 3) = 0$$

So we have roots at $x = 0, \pm 3$. Since the function is a polynomial, we know it has no division by zero and so no vertical asymptotes; similarly as $x \to \pm\infty$ the function diverges, so no horizontal asymptotes. Now we are ready for derivative information.

A quick derivative and we see $f'(x) = x^2 - 3$. So we have critical values at $x = \pm\sqrt{3}$ and critical points at $(\pm\sqrt{3}, \mp 2\sqrt{3})$. Plugging the values $x = 0, \pm 2$ into $f'(x)$ gives us the increasing/decreasing sign chart:

$$(-\infty) + + + (-\sqrt{3}) - - - (\sqrt{3}) + + + (\infty)$$

This tells us the function is increasing on $(-\infty, -\sqrt{3}) \cup (\sqrt{3}, \infty)$ and decreasing on $(-\sqrt{3}, \sqrt{3})$.

Another quick derivative and we see $f''(x) = 2x$. So there is an inflection value at $x = 0$ and an inflection point at $(0,0)$. Plugging in ± 1 to the second derivative yields the sign chart:

$$(-\infty) - - - (0) + + + (\infty)$$

So, the function is concave down on $(-\infty, 0)$ and concave up on $(0, \infty)$.

> **Roots:** $x = 0, \pm 3$
> **Vertical asymptotes:** none
> **Horizontal asymptotes:** none
> **Critical points:** $(\pm\sqrt{3}, \mp 2\sqrt{3})$
> **Increasing on:** $(-\infty, -\sqrt{3}) \cup (\sqrt{3}, \infty)$
> **Decreasing on:** $(-\sqrt{3}, \sqrt{3})$
> **Inflection points:** $(0, 0)$
> **Concave up on:** $(-\infty, 0)$
> **Concave down on:** $(0, \infty)$

The following picture displays all the information. Notice how roots, critical points, and inflection points are all displayed.

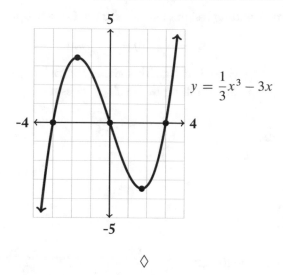

$$y = \frac{1}{3}x^3 - 3x$$

Let's move on to an example with both vertical and horizontal asymptotes.

Example 4.46   Make a full report and sketch for

$$y = \frac{x^2 - 4}{x^2}.$$

Solution:

To find the roots, we solve the numerator equal to zero, $x^2 - 4 = 0$. This gives us roots $x = \pm 2$. The denominator is zero at $x = 0$, giving us a vertical asymptote at $x = 0$. The rule for ratios of polynomials tells us:

$$\lim_{x \to \infty} \frac{x^2 - 4}{x^2} = 1$$

So we have a horizontal asymptote at $y = 1$.
Computing the first derivative we get:

$$f'(x) = \frac{x^2(2x) - (x^2 - 4)2x}{(x^2)^2} = \frac{8x}{x^4} = \frac{8}{x^3}$$

Since the numerator is a constant, there are no critical values and hence no critical points. We build a sign chart on the vertical asymptote, plugging in $f'(\pm 1) = \pm 8$, and get:

$$(-\infty) - - - (0) + + + (\infty)$$

This gives us decreasing on $(-\infty, 0)$ and increasing on $(0, \infty)$. Since $f'(x) = 8x^{-3}$, we can use the power rule to get that $f''(x) = -24x^{-4} = -\dfrac{24}{x^4}$. Again the top is a constant so there are no inflection values or points. Building a sign chart on the vertical asymptote at $x = \pm 1$ we get:

$$(-\infty) --- (0) --- (\infty)$$

In fact $f''(x)$ is negative everywhere it exists. This yields concave up: nowhere; concave down: $(-\infty, 0) \cup (0, \infty)$. And, we are done. Filling in the chart gives us:

Roots: $x = \pm 2$
Vertical asymptotes: x=0
Horizontal asymptotes: y=1
Critical points: none
Increasing on: $(0, \infty)$
Decreasing on: $(-\infty, 0)$
Inflection points: none
Concave up on: never
Concave down on: $(-\infty, 0) \cup (0, \infty)$

The corresponding sketch looks like this:

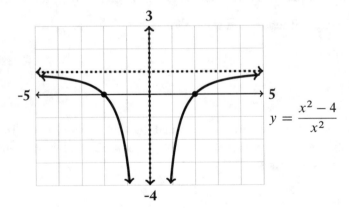

Example 4.47   Make a full report and sketch for

$$y = \tan^{-1}(3x).$$

Solution:

We already know that $\tan^{-1}(x)$ is zero only at $x = 0$. So, solving $3x = 0$ we get a root at $x = 0$. We also know that $-\pi/2 < \tan^{-1}(x) < \pi/2$. So, there are no points where a divide by zero and hence a vertical asymptote can form.

We have that $\lim\limits_{x\to\pm\infty} \tan(x) = \pm\infty$. The quantity $3x$ heads for ∞ faster than x, but it will not change the asymptotes of $y = \pm\pi/2$.

Using the chain rule we see that $f'(x) = \dfrac{3}{1 + 9x^2}$. The numerator is constant, so there are no critical points. The first derivative is the ratio of two positive quantities, and so the function increases everywhere.

Using the reciprocal rule we get that $f''(x) = 3 \cdot \dfrac{-18x}{(1 + 9x^2)^2} = \dfrac{-54x}{(1 + 9x^2)^2}$.

The numerator is zero at $x = 0$, and the denominator is not, yielding a single inflection value of $x = 0$ and an inflection point at $(0, 0)$.

Building a sign chart by plugging in $f''(\pm 1) = \mp 0.27$, we get $(-\infty) + + + (0) - - - (\infty)$. So, the function is concave up on $(-\infty, 0)$ and concave down on $(0, \infty)$. Putting all this in the table yields:

Roots: $x = 0$
Vertical asymptotes: none
Horizontal asymptotes: $y = \pm\pi/2$
Critical points: none
Increasing on: $(-\infty, \infty)$
Decreasing on: never
Inflection points: $(0,0)$
Concave up on: $(-\infty, 0)$
Concave down on: $(0, \infty)$

The resulting graph is:

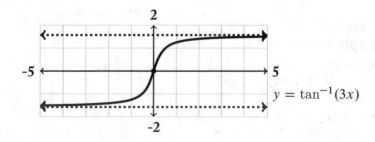

$$y = \tan^{-1}(3x)$$

$\Diamond$

Example 4.48   Make a full report and sketch for

$$y = \frac{x^2 - 9}{x}.$$

Solution:

The roots are easy, $x = \pm 3$; the vertical asymptote is clearly $x = 0$; and, since the top is higher degree, there are no horizontal asymptotes. Notice that:

$$y = \frac{x^2 - 9}{x} = \frac{x^2}{x} - \frac{9}{x} = x - \frac{9}{x}$$

From this, it is easy to see that $f'(x) = 1 + \dfrac{9}{x^2}$, which means there are no critical points, and $f'(x) > 0$ where it exists. This makes the first derivative sign chart

$$(-\infty) + + + (0) + + + (\infty)$$

The second derivative is $f''(x) = \dfrac{-18}{x^3}$, so there are no inflection points. Plugging in $f''(\pm 1) = \mp 18$, we get a second derivative sign chart like this:

$$(-\infty) + + + (0) - - - (\infty)$$

These sign charts make the various ranges obvious and we get:

Roots: $x = \pm 3$
Vertical asymptotes: x=0
Horizontal asymptotes: none
Critical points: none
Increasing on: $(-\infty, 0) \cup (0, \infty)$
Decreasing on: never
Inflection points: none
Concave up on: $(-\infty, 0)$
Concave down on: $(0, \infty)$
The corresponding sketch looks like this:

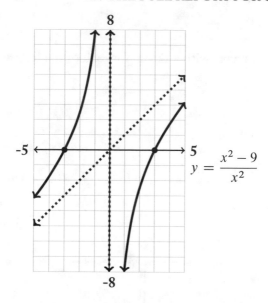

$$y = \frac{x^2 - 9}{x^2}$$

Alert students will notice that a non-standard asymptote appears in this picture. Recall that $y = x - \frac{9}{x}$. When $|x|$ is large, $y = x -$ something very small, which means that as $|x|$ grows large, the function gets very close to the line $y = x$. This is called a **diagonal asymptote**.

Knowledge Box 4.10

Diagonal asymptotes

If $f(x) = ax + b + g(x)$ and

$$\lim_{x \to \pm\infty} g(x) = 0,$$

then the graph of $f(x)$ approaches the line $y = ax + b$ as $|x|$ gets large.

PROBLEMS

Problem 4.49   Make a full report with a sketch for each of the following functions.

1. $f(x) = x^3 - 5x$

2. $g(x) = x^3 - 6x^2 + 11x - 6$

3. $h(x) = \dfrac{x^3 - 2x^2 - x + 2}{x}$

4. $r(x) = \dfrac{x^2 - 9}{x^2 - 1}$

5. $s(x) = x + \dfrac{1}{x - 1} + \dfrac{1}{x + 1}$

6. $q(x) = \ln(x^2 + 5)$

7. $a(x) = \tan^{-1}(x^2)$

8. $b(x) = \dfrac{x^2 - 1}{3 - x^2}$

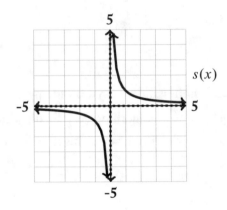

Problem 4.50   Based on the sketch above, do your best to fill out a full report.

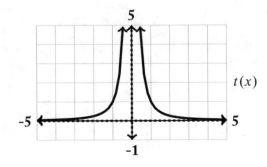

Problem 4.51   Based on the sketch above, do your best to fill out a full report.

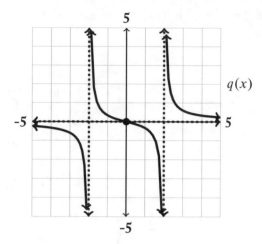

Problem 4.52 Based on the sketch above, do your best to fill out a full report.

Problem 4.53 Make a full report with a sketch for:

$$f(x) = \frac{3x}{x^2 + 1}$$

Problem 4.54 Find a function that is concave down everywhere that it exists and has a range of $(-\infty, \infty)$.

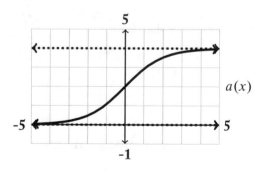

Problem 4.55 Based on the sketch above, do your best to fill out a full report.

Problem 4.56   Make a full report with a sketch for each of the following functions. Include diagonal asymptotes if any.

1. $f(x) = \dfrac{x^2 + 2x + 1}{x - 1}$

2. $g(x) = \dfrac{2x^2 + 3x + 1}{x}$

3. $h(x) = 5x - 3 + \dfrac{1}{x^2}$

4. $r(x) = \dfrac{x^3}{x^2 + 2}$

5. $s(x) = \dfrac{x^2}{x^2 + 1}$

6. $q(x) = \dfrac{x^2}{x^2 - 4}$

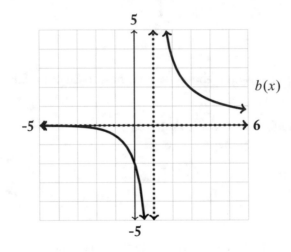

Problem 4.57   Based on the sketch above, do your best to fill out a full report.

Problem 4.58   Make a full report with a sketch for

$$f(x) = \frac{e^x}{1 + e^x}$$

Problem 4.59   Find a function that has two vertical asymptotes and approaches the diagonal asymptote $y = 2x + 1$.

Problem 4.60   Find a function with two horizontal and two vertical asymptotes and demonstrate that your solution is correct.

Problem 4.61   Suppose that a function is the ratio of two polynomials in which the numerator has degree n and the denominator has degree m. What is the range of the possible numbers of vertical asymptotes of this function?

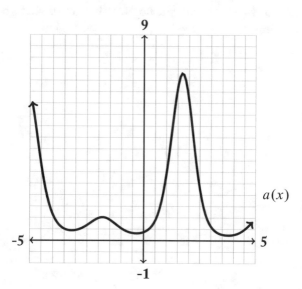

Problem 4.62   Based on the sketch above, do your best to fill out a full report.

Problem 4.63   Find a function that has two horizontal asymptotes and explain why your solution works.

Problem 4.64   Find a function that has two diagonal asymptotes and explain why your solution is correct. Hint: start with a function that has two horizontal asymptotes.

Problem 4.65   What is the relationship between the graph of

$$f(x) = \frac{x^4}{x^2 + 1}$$

and the graph of $y = x^2 - 1$?

Problem 4.66 Suppose that

$$f(x) = \frac{(x-a)(x-b)(x-c)}{x^4 + d}$$

where all four of a, b, c, and d are positive constants. What items that go into a report can you deduce from this information?

CHAPTER 5

Optimization

In this chapter we will start on **optimization**, the process of finding the largest or smallest value a function can take on as well as values that are larger (or smaller) than all nearby values. This chapter only lets us optimize functions of one independent variable, but it sets the stage for more complex types of optimization later in the series.

5.1   OPTIMIZATION WITH DERIVATIVES

As is often the case, our first step is to define our terms. We need terminology to sort out different types of maximum and minimum value.

Definition 5.1 *If a function $f(x)$ is defined on an interval or collection of intervals $\mathcal{I}$, then the* **global maximum** *of $f(x)$ is the largest value $f(x)$ takes on anywhere on $\mathcal{I}$.*

Definition 5.2 *If a function $f(x)$ is defined on an interval or collection of intervals $\mathcal{I}$, then the* **global minimum** *of $f(x)$ is the smallest value $f(x)$ takes on anywhere on $\mathcal{I}$.*

Definition 5.3 *If a function $f(x)$ is defined near and including $x = c$ and, in some interval $I = (c - \delta, c + \delta)$, we have that, for all $a \in I$, $f(c) \geq f(a)$, then we say $f(x)$ has a* **local maximum** *at $x = c$.*

Definition 5.4 *If a function $f(x)$ is defined near and including $x = c$ and, in some interval $I = (c - \delta, c + \delta)$, we have that, for all $a \in I$, $f(c) \leq f(a)$, then we say $f(x)$ has a* **local minimum** *at $x = c$.*

We use the term **optima** for values that are maxima or minima. Figure 5.1 demonstrates the different sorts of optima.

The key to finding optima is the behavior of the derivative. Notice that, at an optimum, the function changes between increasing and decreasing. This means that optima, at least ones that don't occur at the beginning and end of an interval, are a kind of point we've seen before: they

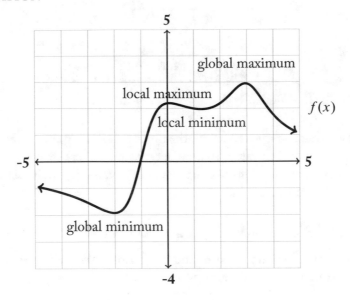

Figure 5.1: This picture shows examples of local and global optima.

are critical points.

Knowledge Box 5.1

Critical points in optimization

Except at the boundaries of an optimization problem, the optima of a continuous, differentiable function occur at critical points.

If we consider the tangent line at a critical point, we get a geometric description of the behavior of the derivative. A line with a slope of 0 is a horizontal tangent line. This tells us that **optima can occur at horizontal tangent lines**.

A function that diverges to infinity as $|x|$ gets large may fail to have a global maximum or minimum. Similarly, vertical asymptotes inside the area where we are optimizing can cause trouble. Having said that, the other place besides critical points where optima can occur is at the boundaries of the region where we are optimizing. In general, when optimizing a continuous, differentiable function, we check the critical points and the points at the boundaries of the area we are optimizing.

The next example asks you to optimize a quadratic equation—something that should be easy—but it asks you to do it on a bounded domain. A quadratic that opens upward has a unique minimum. As well as the critical points, we must also check the x values at its endpoints.

Example 5.1   Find the global maximum and minimum of $g(x) = x^2 - 4$ on the interval $[-2,3]$.

Solution:

The critical point of this function is easy: $g'(x) = 2x$. So $x = 0$ is the critical value, and there is a critical point at $(0,-4)$. The boundaries of the optimization area are $x = -2$ and $x = 3$. Since $g(-2) = 0$ and $g(3) = 5$, we have three candidate points: $(-2,0)$, $(0,-4)$, and $(3,5)$. The smallest y-value is -4 making $(0,-4)$ the point where the global minimum occurs, while 5 is the largest y-value making $(3,5)$ the point where the global maximum occurs. Let's look at a picture.

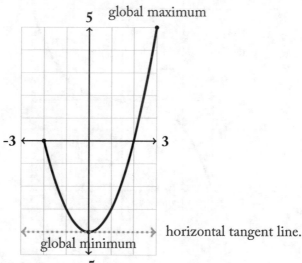

The function $g(x) = x^2 - 4$ on the interval $[-2,3]$

$\Diamond$

Knowledge Box 5.2 summarizes what we have so far.

Knowledge Box 5.2

The Extreme Value Theorem

The global maximum and minimum of a continuous, differentiable function must occur at critical points (horizontal tangents) or at the boundaries of the domain where optimization is taking place.

An important point is that critical points don't have to be maxima or minima. Let's look at an example of this. Consider the function

$$h(x) = \frac{x^3}{9},$$

shown in Figure 5.2.

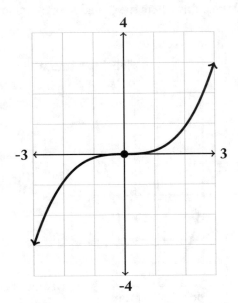

Figure 5.2: The function $h(x) = \dfrac{x^3}{9}$.

Since $f'(x) = \frac{1}{3}x^2$, it's easy to see there is a critical point at $(0, 0)$—but it's *not* a maximum or a minimum. This then opens the question, **how do we tell if a critical point is a maximum, a minimum, or neither?** We have already developed one useful tool: the sign chart for the first derivative. Let's look at the sign chart for $h(x) = \dfrac{x^3}{9}$:

$$(-\infty) + + + (0) + + + (\infty)$$

The function increases to zero and then continues to increase. There are four ways a sign chart can shake out which together are the **first derivative test** for the nature of an optimum.

Knowledge Box 5.3

The first derivative test for optima

$f'(x)$ chart near p	Conclusion:
$+++(p)+++$	no optimum at p
$+++(p)---$	maximum at p
$---(p)+++$	minimum at p
$---(p)---$	no optimum at p

Another method of spotting a potential maximum or minimum is to look at the second derivative. A function that is curved downward at a critical point has a maximum at that critical point; a function that is curved upward at a critical point has a minimum at that critical point. This technique is called the **second derivative test**.

Knowledge Box 5.4

The second derivative test for optima

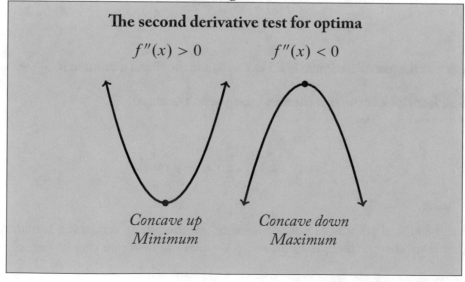

$f''(x) > 0$ $f''(x) < 0$

Concave up   *Concave down*
Minimum   *Maximum*

This gives us the tools we need to optimize things. Now it is time to do some examples.

Example 5.2   Find the global maximum and minimum of

$$f(x) = \frac{1}{2}x^3 - 2x$$

on the interval $[-1, 3]$.

Solution:

First we find the critical points. Solving $f'(x) = \frac{3}{2}x^2 - 2 = 0$ we get critical values of $\pm\frac{2}{\sqrt{3}}$, but only the positive value is in the interval $[-1,3]$. This means we need to check this value and the ends of the interval:

$$f(-1) = 3/2 = 1.5$$
$$f\left(\frac{2}{\sqrt{3}}\right) = -\frac{8}{3\sqrt{3}} \cong -1.54$$
$$f(3) = 15/2 = 7.5$$

This means that the global maximum is 7.5 at $x = 3$, and the global minimum is $-\frac{8}{3\sqrt{3}} \cong -1.54$ at $x = \frac{2}{\sqrt{3}}$. Let's look at the sign chart and the graph. The chart:

$$(-1) - - - (\frac{2}{\sqrt{3}}) + + + (3)$$

shows that the critical point is, in fact, a minimum. The maximum occurs at a boundary point. Notice that if we change the interval on which we are optimizing we can, in fact, change the results.

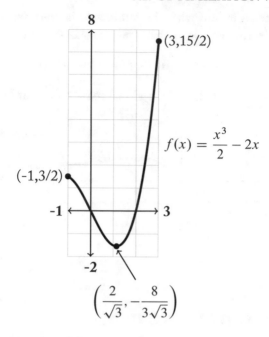

We can also look at the outcome of the second derivative test.

$$f''(x) = 3x$$

So if we plug in $x = \dfrac{2}{\sqrt{3}}$ we get a positive value, about 3.46. This means that the graph is concave up and so, again, we have determined that there is a minimum at the critical point.

Example 5.3   Find the global optima of

$$g(x) = \frac{x}{1+x^2}.$$

Solution:

Since no interval is given, we use the interval $(-\infty, \infty)$. Our rule about the ratio of polynomials tells us that the limits of this function at $\pm\infty$ are zero. So the optima occur at critical points, if they occur. Using the quotient rule,

$$g'(x) = \frac{(1+x^2)(1) - x(2x)}{(1+x^2)^2} = \frac{1-x^2}{(1+x^2)^2}$$

Remembering that a fraction is zero when its numerator is zero, we get critical values where $x^2 - 1 = 0$ or $x = \pm 1$. Examining a sign chart, plugging in $0, \pm 2$ to $f'(x)$ we get

$$(-\infty) - - - (-1) + + + (1) - - - (\infty)$$

So the global minimum is at (-1,-1/2) and the global maximum is at (1,1/2). The graph is:

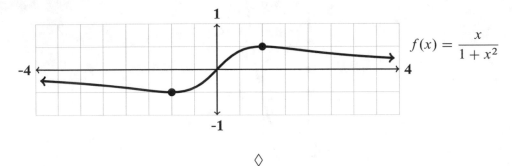

$$f(x) = \frac{x}{1 + x^2}$$

$\Diamond$

5.1.1   OPTIMIZATION STORY PROBLEMS

We now turn to optimization story problems. These problems describe a situation and ask the reader to optimize some quantity. Common tasks are designing containers and laying out fences, but there are actually a huge number of possible story problems.

Example 5.4   Suppose we have 120 m of fence that we want to use to enclose two pens that share a wall as shown. If one of the pens is $a \times b$ meters, what values of a and b maximize the area of the pens? What is the area of the optimal pen?

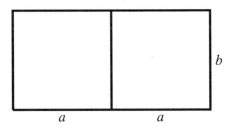

Solution:

There are four segments of length a and three segments of length b, so we have $4a + 3b = 120$. We want to maximize the area of the pens, Area$(a, b) = ab$. Since $b = (120 - 4a)/3$ we can maximize

$$\text{Area} = \frac{a(120 - 4a)}{3} = \frac{120}{3}a - \frac{4}{3}a^2$$

Take a derivative and we see that Area$' = \dfrac{120}{3} - \dfrac{8}{3}a$. Solving Area$' = 0$, we obtain a critical value of $a = \dfrac{120}{8} = 15$ m, and so $b = \dfrac{120}{6} = 20$ m.

The area function is a quadratic with a negative squared term that opens downward—so the single critical point is the maximum.

The answer is thus: $a = 15$ m, $b = 20$ m, and Area $= 300$ m^2.

Knowledge Box 5.5

Steps for an optimization story problem

- *Draw a picture of the problem*

- *Label the picture with reasonable variables*

- *Write out the quantity you are optimizing in terms of those variables*

- *Write out the equation for additional information*

- *Use the additional information to remove a variable from the quantity you are optimizing*

- *Figure out what the interval for that variable is, based on the problem*

- *Optimize the resulting single variable formula*

It may help to match these steps against the previous example.

Example 5.5   Suppose that we have an open-topped, square-bottomed box with a volume of 1m^3. What side length s for the base and height h minimize the amount of material needed to make the box?

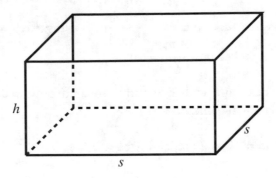

Solution:

We are minimizing the surface area A of the box to minimize the material. The surface area is the square bottom and four rectangular sides, making $A = s^2 + 4sh$. We also know the volume of the box is 1 m^3, so $1 = s^2h$ or $h = \dfrac{1}{s^2}$. Substituting this value for h into the area formula we get

$$A = s^2 + 4s\frac{1}{s^2} = s^2 + \frac{4}{s}$$

We know $s > 0$, so the interval is $(0, \infty)$. Taking the derivative we get

$$A' = 2s - \frac{4}{s^2} = \frac{2s^3 - 4}{s^2}$$

Solving this for zero, we get that $2s^3 - 4 = 0$ or that $s = \sqrt[3]{2}$ is the sole critical value. The second derivative is easy to compute:

$$A'' = 2 + \frac{8}{s^2}$$

This function is positive for any positive s. So we get $A''(\sqrt[3]{2}) > 0$, meaning that the critical value is a minimum. This means we have $s = 2^{1/3} \cong 1.26$ m and, plugging into the formula for h, $h = 2^{-2/3} \cong 0.630$ m. These are the side length and height that minimize materials.

Example 5.6   Suppose we are fencing in a rectangular pen with 200 m of fence, but we are putting the pen against the side of an existing building. What are the dimensions that enclose the maximum area?

Solution:

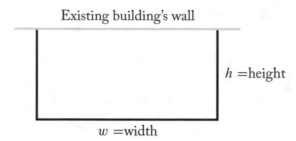

Existing building's wall

$h =$ height

$w =$ width

We are optimizing area $A = wh$, and we know that $w + 2h = 200$. So $w = 200 - 2h$. Substituting this into our area formula we get:

$$A = (200 - 2h)h = 200h - 2h^2$$

This function is a quadratic that opens downward. So we know that it will have a single critical value that corresponds to a maximum. Taking the derivative $A' = 200 - 4h$ gives us a critical value of $h = 50$, meaning $w + 100 = 200$ or $w = 100$, and we have height $= 50$ m and width=100m giving us the best possible area for the pen.

An important point is that some functions don't have optima. The next example illustrates this.

Example 5.7   Find the global optima, if any, of $y = e^x$.

Solution:

We know that as $x \to \infty$ this function diverges to ∞, so there is no hope that there is a global maximum. As $x \to -\infty$ the function goes to zero. What, however, is the range of the function? We already know that it is $(0, \infty)$. This means that $y = e^x$ can take on every positive value, but it can never be zero. This means that the function does not have a global minimum. For any value that e^a takes on, there is some b so that $e^b < e^a$.

This is a fairly weird way to fail to have an optimum. It's related to the strict inequality

$$0 < e^x < \infty$$

that defines the range of $y = e^x$. This sort of thing can also happen when the interval over which a function is optimized has open ends, (a, b) instead of closed ones $[a, b]$.

Example 5.8   What radius and height minimize the material needed to make a can with a volume of 200 cc?

Solution:

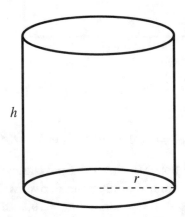

Again, assume we are minimizing surface area. The area of the can is the two circular ends with area πr^2 and the sides which, if flattened, form a $2\pi r \times h$ rectangle. So

$$A = 2\pi r^2 + 2\pi r h$$

The volume of a can is the area of the bottom times the height, which tells us

$$V = \pi r^2 h = 200$$

From this we deduce that $h = \dfrac{200}{\pi r^2}$. Plugging this in to get a single-variable problem, we get

$$A = 2\pi r^2 + 2\pi r \cdot \frac{200}{\pi r^2} = 2\pi r^2 + \frac{400}{r}$$

Setting the derivative equal to zero gives us:

$$4\pi r - \frac{400}{r^2} = 0$$
$$\frac{4\pi r^3 - 400}{r^2} = 0$$
$$4\pi r^3 - 400 = 0$$
$$r^3 = \frac{100}{\pi}$$
$$r \cong 3.17 \text{ cm}$$

We know $r > 0$. Looking at the sign chart for the derivative, plugging in $r = 2$ and $r = 4$, we get:

$$(0) - - - (3.17) + + + (\infty)$$

so the critical value $x = 3.17$ is a minimum. Plugging the r value into the formula for h, we get that $h \cong 6.34$. Our answer is $r = 3.17$ cm and $h = 6.34$ cm.

$$\Diamond$$

Let's see how the second derivative test shakes out in the previous example.

$$A'' = 4\pi + \frac{800}{r^3}$$

which is positive for any $r > 0$. So the curve is concave up in the possible region, and our critical value is a minimum; the second derivative test agrees with the sign chart for the first derivative test.

When you are working optimization story problems, it is critical to make sure the values you get make sense. Negative lengths, for example, probably mean you made a mistake. These problems also have the property that some of what you did may be needed again in a later step. Using a neat layout, possibly informed by the steps given in Knowledge Box 5.5, will help.

PROBLEMS

Problem 5.9   For each of the following functions, find the global maximum and minimum of the function, if they exist, on the stated interval.

1. $f(x) = 2x - 1$ on $(-2, 2)$

2. $g(x) = 3x + 1$ on $[-3, 1]$

3. $h(x) = x^2 + 4x + 3$ on $[-1, 4]$

4. $r(x) = \ln(x)$ on $[1, e^3]$

5. $s(x) = e^{-x^2}$, on $[-2, 3]$

6. $q(x) = x^3 - 16x + 1$, on $[-4, 4]$

Problem 5.10   How many different horizontal tangent lines do the following functions have? Be sure to justify your answer.

1. $f(x) = \cos(x)$

2. $g(x) = xe^{-x}$

3. $h(x) = x(x-5)(x+5)$

4. $r(x) = \cos(x) + x$

5. $s(x) = \ln(x^4 + 4x^3 + 5)$

6. $q(x) = \tan^{-1}(x)$

Problem 5.11   Construct a function that has exactly three horizontal tangent lines—all different from one another.

Problem 5.12   What is the largest number of horizontal tangent lines that a polynomial of degree n can have?

Problem 5.13   If $y = ax^2 + bx + c$ with $a \neq 0$, give a set of steps for finding the global optimum (there is exactly one), and determining the type of optima it is.

Problem 5.14   For each of the following functions, find the global maximum and minimum of the function, if they exist, on the stated interval.

1. $f(x) = \dfrac{4x}{x^2 + 2}$ on $(-\infty, \infty)$

2. $g(x) = e^{x^3 - 5x + 12}$ on $[-5 : 5]$

3. $h(x) = xe^{-2x}$ on $(0, \infty)$

4. $r(x) = x^2 e^{-x}$ on $(0, \infty)$

5. $s(x) = \dfrac{\sqrt{x^2 + 1}}{x}$, on $[-4, -1] \cup [1, 4]$

6. $q(x) = 25 - x^4$, on $[-1, 2]$

Problem 5.15   Suppose for $m(x)$ that when $a \leq b$ we have that $m(a) \leq m(b)$, and assume that $m(x)$ is continuous, differentiable, and not constant.

1. What do we know about $m'(x)$? Explain.

2. Prove that the critical values of $m(f(x))$ and $f(x)$ are the same.

3. Is this a problem relevant to other problems in this section? Why?

Problem 5.16   Find the domain of

$$h(x) = \sqrt{6 - 3x - x^2}$$

and find its global maximum and minimum on the domain.

Problem 5.17   If an open-topped can holds 400 cc, what radius and height minimize the amount of material needed to make the can?

Problem 5.18   What are the ratio of the sides of a rectangle of perimeter P that maximizes the area over all rectangles with that perimeter?

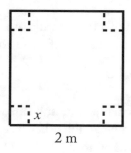

2 m

Problem 5.19   Suppose that we cut square corners out of a 2 m × 2 m square of pasteboard and tape up the sides to make an open-topped box. What side length x of the square maximizes the volume of the box?

Problem 5.20   What point on the line

$$y = x - 6$$

is closest to the origin?

Problem 5.21   What point on the line

$$y = 3x + 5$$

is closest to the origin?

Problem 5.22   What point on the line

$$y = 4x + 1$$

is closest to the origin?

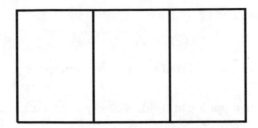

Problem 5.23   If we use 240 m of fence to lay out three pens like those shown above, what length and width *for one pen* maximize the area enclosed?

Problem 5.24   Suppose that the top of a can is made of a material that costs twice as much as the bottom or sides. Find the radius and height of a can containing 200 cc that minimizes the cost.

Problem 5.25   One side of a rectangular pen must be made of opaque material that costs three times as much as the material used to make the other three sides. What dimensions minimize the cost if the pen must have an area of 60 m^2?

Problem 5.26   A rocket takes off, straight up, so that for $t \geq 0$ the height of the rocket is $\dfrac{4t^2}{m}$. A camera 40 m from the rocket tracks its takeoff.

1. Make a sketch of the situation.

2. Find an expression for the rate in rad/sec for the camera's rate of spin.

3. When is the camera spinning fastest?

4. What is the camera's rate of spin when it is spinning fastest?

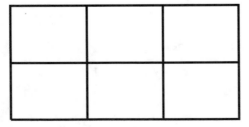

Problem 5.27   Suppose we are optimizing the area of cells of a rectangular grid like the one shown above fixing the total length of the sides of the grids. Prove that the solution always places the same amount of material into vertical and horizontal cell sides.

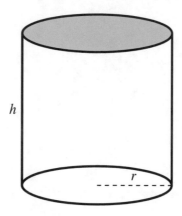

Problem 5.28   What radius and height minimize the material needed to make an open-topped can with a volume of 600 cc?

Problem 5.29   Maximize

$$q(x) = x^2 e^{-2x}$$

for $x \geq 0$.

Problem 5.30   What is the global maximum of the function

$$y = e^{4-x^2}$$

Can you do this problem without calculus? Explain.

CHAPTER 6

Limits and Continuity: The Details

In this chapter we go into the material usually presented at the beginning of calculus courses—the formal definition of limits and continuity. While we've taught you a boat-load of tricks for doing limits, we've largely ignored continuity (other than to say you need it) and the limits, so far, are all rules of thumb. Continuity is based on limits so we do them first.

6.1   LIMITS AND CONTINUITY

If $\lim_{x \to a} f(x) = L$ then, informally, as x gets closer to a, $f(x)$ needs to get closer to L. The absolute value function gives us the ability to figure out if two things are "close." If $|x - a|$ is small, x is close to a; if $|f(x) - L|$ is small, then $f(x)$ is close to L. We have two traditional symbols to represent small values: the Greek letters epsilon, ϵ, and delta, δ. The closeness of x to a is encoded with δ, while the closeness of $f(x)$ to L is encoded by ϵ. With that preparation, here is the definition:

Knowledge Box 6.1

The formal definition of a limit

Let $f(x)$ be a function defined on an open interval containing the value $x = a$, except that $f(x)$ might not be defined at a itself. We say that the limit of $f(x)$ at a is L, written

$$\lim_{x \to a} f(x) = L$$

if, for every $\epsilon > 0$, it is possible to find $\delta > 0$ so that

if $0 < |x - a| < \delta$, then we have that $|f(x) - L| < \epsilon$.

Example 6.1   Show formally that

$$\lim_{x \to 1} (2x + 1) = 3.$$

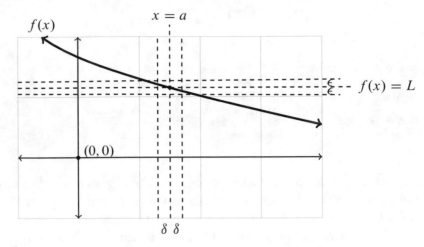

Figure 6.1: The geometry of $\lim\limits_{x \to a} f(x) = L$. As long as x is within δ of a, we have that $f(x)$ is within ϵ of L.

Solution:

Since there are no algebraic anomalies, we can just plug in and see that the desired limit is actually 3.

But, how do we show it using the formal definition?

We have two inequalities to satisfy. For one, we are given a positive number ϵ; for the other we need to find an appropriate positive number δ. Let's start with the inequality where everything is known:

$$|f(x) - L| < \epsilon$$

$$|2x + 1 - 3| < \epsilon$$

$$|2x - 2| < \epsilon - \epsilon \qquad\qquad < 2x - 2 < \epsilon$$

$$-\epsilon < 2(x - 1) < \epsilon$$

$$-\frac{\epsilon}{2} < x - 1 < \frac{\epsilon}{2}$$

$$|x - 1| < \frac{\epsilon}{2}$$

This gives us information about how to choose δ. Since $a = 1$, we need to find a δ such that $|x - 1| < \delta$. So, set $\delta = \frac{\epsilon}{2}$. Then we satisfy the formal definition, as shown below.

Set $\delta = \frac{\epsilon}{2}$ and suppose that $0 < |x - 1| < \delta$. Then

$$|x - a| < \delta \qquad \text{The start of the formal definition.}$$

$$|x - 1| < \delta \qquad \text{Substitute in actual values.}$$

$$|x - 1| < \frac{\epsilon}{2}$$

$$|2x - 2| < \epsilon \qquad \text{Multiply by 2}$$

$$|2x + 1 - 3| < \epsilon$$

$$|f(x) - L| < \epsilon \qquad \text{Conclusion of the formal definition.}$$

So now we have proven, formally, that the limit is 3.

$$\diamond$$

Lines are the simple case—for any more complex functions, the algebra gets more complicated.

Example 6.2 Show formally that

$$\lim_{x \to 4} x^2 = 16$$

Solution:

Our goal is to have

$$\text{If } 0 < |x - 4| < \delta \text{ then } |x^2 - 16| < \epsilon$$

We will use the fact that when x is near 4 then $x + 4 < 9$. Compute:

$$|x - 4| < \delta$$

$$|x - 4| \cdot |x + 4| < |x + 4|\delta$$

$$|x^2 - 16| < |x + 4|\delta < 9\delta$$

$$|x^2 - 16| < 9\delta$$

So let us set $\delta = \epsilon/9$. Check the formal definition.

$|x - a| < \delta$	Start of the formal definition.
$|x - 4| < \delta$	
$|x - 4||x + 4| < |x + 4|\delta$	
$|x^2 - 16| < |x + 4|\delta < 9\delta$	Recall: $|x + 4| < 9$
$|x^2 - 16| < 9\delta$	
$|x^2 - 16| < \epsilon$	
$|f(x) - L| < \epsilon$	Conclusion of the formal definition.

And we have satisfied the formal definition.

Notice that the formulas we are using that tell us how ϵ and δ are related have the property that as δ shrinks to zero, so does ϵ.

Like derivatives and integrals, limits have some useful algebraic properties.

Knowledge Box 6.2

Algebraic properties of limits

Suppose that $\lim\limits_{x \to a} f(x) = L$, $\lim\limits_{x \to a} g(x) = M$, *and that c is a constant.*
Then:

- $\lim\limits_{x \to a} (f(x) \pm g(x)) = L \pm M$

- $\lim\limits_{x \to a} c \cdot f(x) = c \cdot L$

If one needs to take a problematic limit, these properties permit you to break it up into pieces. Also, since derivatives are based on limits, these algebraic roles underlie the summing and scaling properties of derivatives.

In Chapter 3 we informally defined limits of a function from above and below a value $x = c$. The formal definitions of these limits are closely parallel to the formal definition of a limit.

Knowledge Box 6.3

The formal definition of limits from above and below

We say that

$$\lim_{x \to a^-} f(x) = L$$

if, for each $\epsilon > 0$, we can find $\delta > 0$ so that

$$\textit{If } a - \delta < x < a, \text{ then } |f(x) - L| < \epsilon.$$

Similarly, we say that

$$\lim_{x \to a^+} f(x) = L$$

if, for each $\epsilon > 0$, we can find $\delta > 0$ so that

$$\textit{If } a < x < a + \delta, \text{ then } |f(x) - L| < \epsilon.$$

The following example needs a one sided limit, because the function we are taking the limit of exists only on one side of the limiting point.

Example 6.3   Prove that

$$\lim_{x \to 0^+} \sqrt{x} = 0$$

Solution:

When we have that $0 < x < \delta$ we need for $|\sqrt{x} - 0| < \epsilon$, which is equivalent to $\sqrt{x} < \epsilon$. Squaring both sides we get $x < \epsilon^2$. Choose $\delta = \epsilon^2$ and check the definition.

$a < x < a + \delta$ Beginning of formal definition

$0 < x < \delta$

$0 < x < \epsilon^2$

$0 < \sqrt{x} < \epsilon$

$0 < \sqrt{x} - 0 < \epsilon$

$|\sqrt{x} - 0| < \epsilon$ Conclusion of the formal definition

At this point, for completeness, we also restate the definition of limits at infinity. We originally encountered these in Chapter 4.

Knowledge Box 6.4

The formal definition of a limit at infinity

If we can find a number L so that, for every number $\epsilon > 0$, we can find a value X_ϵ such that, when $x \geq X_\epsilon$, we have that $|f(x) - L| < \epsilon$, then L is the limit of $f(x)$ as $x \to \infty$. Written:

$$\lim_{x \to \infty} f(x) = L$$

Example 6.4 Using the definition of the limit, show that

$$\lim_{x \to \infty} \frac{1}{3 + x^2} = 0$$

Solution:

The trick here is to find an appropriate X_ϵ, given an ϵ. Remember that when you take the reciprocal of both sides of an inequality, it reverses.

$$\left| \frac{1}{3 + X_\epsilon^2} - 0 \right| = \frac{1}{3 + X_\epsilon^2} < \epsilon$$

$$3 + X_\epsilon^2 > \frac{1}{\epsilon}$$

$$X_\epsilon^2 > \frac{1}{\epsilon} - 3$$

$$X_\epsilon > \sqrt{\frac{1}{\epsilon} - 3}$$

So the value

$$X_\epsilon = \sqrt{\frac{1}{\epsilon} - 3}$$

meets the requirements if $\epsilon \leq 1/3$.

If $\epsilon > 1/3$, then this value is undefined. So, in that case, just pick $X_\epsilon = 0$. $f(0) = 1/3$ and f is decreasing. So, when $x \geq 0$, $f(x) \leq 1/3 < \epsilon$ for $\epsilon > 1/3$.

$\Diamond$

We also remind the reader of the formal definition of diverging to infinity from Chapter 4.

Knowledge Box 6.5

The formal definition of divergence to ∞

If, for any constant c, we can find X_c so that $x > X_c$ means $f(x) \geq c$, then

$$\lim_{x \to \infty} f(x)$$

diverges to infinity.

Example 6.5   Show formally that

$$\lim_{x \to \infty} x^4 = \infty.$$

Solution:

For $f(x) = x^4$ to be bigger than c we need $x > \sqrt[4]{c}$. So the choice of X_c is pretty simple. Let $X_c = \sqrt[4]{c}$ and check the formal definition.

Suppose that $x > X_c = \sqrt[4]{c}$, then $f(x) \geq (\sqrt[4]{c})^4 = c$, and we meet the requirements of the formal definition.

$$\Diamond$$

Once we have limits, we can define continuity. Suppose that a function $f(x)$ exists on an interval $a < x < b$. Then, informally, $f(x)$ is continuous on an interval if you can draw the graph of the function on $a < x < b$ *without lifting your pencil*. (Mathematics should not, of course, be done in pen unless required for some formality of testing.) The formal definition is more complex.

Knowledge Box 6.6

The definition of continuity at a point

A function $f(x)$ is **continuous at x=c** *if three things are true.*

1. $\lim_{x \to c^-} f(x) = v,$

2. $\lim_{x \to c^+} f(x) = v,$ *and*

3. $f(c) = v.$

In other words, $f(x)$ is continuous at $x = c$ if the limits from above and below exist and are equal to the value of the function at $x = c$.

The definition of continuity at a point can fail in a number of ways. A function could lack one or both of the required limits; the limits could disagree with one another or the value of the function; or the limits might be fine but the function might not exist as with the somewhat contrived example

$$f(x) = \frac{x^2 - 4}{x - 2}$$

from Chapter 4.

Example 6.6   Is the function

$$f(x) = \begin{cases} x^2 & x < 2 \\ 3x - 2 & x \geq 2 \end{cases}$$

continuous at $x = 2$?

Solution:

First compute $f(2) = 3 \cdot 2 - 2 = 4$. The limit from below takes place where the rule of the function is x^2. So

$$\lim_{x \to 2^-} f(x) = 4$$

by substituting into x^2. The limit from above takes place where the rule of the function is $3x - 2$. So

$$\lim_{x \to 2^+} f(x) = 4$$

by substituting into $3x - 2$. The function exists with a value of 4 at $x = 2$ which agrees with the limits from above and below. So the function is continuous at $x = 2$.

Example 6.7   Is

$$g(x) = \sqrt{x}$$

continuous at $x = 0$?

Solution:

The function is not continuous at $x = 0$ because the limit from below is in an area where the function does not exist.

◊

The informal definition of continuity said that a function is continuous on an interval if you can draw the graph without lifting your pencil. The formal definition, so far, talks only of continuity at a point. Knowledge Box 6.7 provides the formal definition of continuity on an interval.

Knowledge Box 6.7

The definition of continuity on an interval

A function is continuous on an interval if it is continuous at every point in that interval.

Which is a bit of an anticlimax. Mathematicians have already checked the continuity of a large number of functions and worked out some rules that will permit you to check if functions are continuous without laborious reference to the formal definitions. The first group are simply continuous everywhere.

Knowledge Box 6.8

Functions that are continuous on $-\infty < x < \infty$

- *polynomials,*

- $\sin(x)$, $\cos(x)$, e^x, $\tan^{-1}(x)$,

- $\sqrt[n]{x}$, *when n is odd.*

The next list is not quite *as* nice because the functions have vertical asymptotes or fail to exist for some ranges of x.

Knowledge Box 6.9

Functions that are continuous on intervals where they exist

- *rational functions,*

- $\tan(x)$, $\cot(x)$, $\sec(x)$, $\csc(x)$, $\ln(x)$, $\sin^{-1}(x)$, $\sec^{-1}(x)$,

- $\sqrt[n]{x}$, *when n is even.*

As you would expect, we have algebraic rules for making more continuous functions. In essence, doing standard arithmetic on continuous functions yields a continuous function unless you do something impossible like divide by zero or take an even root of a negative number. If we know the domain and range of two functions, then we may also be able to compose continuous functions to obtain a continuous function.

Knowledge Box 6.10

Operations that preserve continuity

Suppose that $f(x)$ and $g(x)$ are functions that are continuous on the interval $a < x < b$ and that c is a constant. Then the following are also continuous on the interval $a < x < b$.

- $f(x) \pm g(x)$

- $f(x) \cdot g(x)$

- $c \cdot f(x)$

- $f(x)/g(x)$, *if $g(x)$ has no roots r with $a < r < b$*

Example 6.8   Show that $f(x) = \dfrac{\sin(x)}{x^2 + 1}$ is continuous everywhere.

Solution:

We have $\sin(x)$ and $x^2 + 1$ on our list of functions that are continuous everywhere. Since $x^2 \geq 0$, we know that $x^2 + 1 > 0$ so the rule for dividing functions says that $f(x)$ is continuous everywhere, as $x^2 + 1$ has no roots on $-\infty < x < \infty$.

Example 6.9   Determine the largest intervals on which

$$g(x) = \frac{1}{x}$$

is continuous.

Solution:

The function $g(x)$ is a rational function. It exists when $x \neq 0$ so it is continuous on any interval not containing zero. This means the largest intervals where it is continuous are

$$-\infty < x < 0 \text{ and } 0 < x < \infty$$

◊

Knowledge Box 6.11

Composition of continuous functions

Suppose that $g(x)$ is continuous on $a < x < b$, that on the interval $a < x < b$ all values of $g(x)$ are in the interval $c < x < d$, and that $f(x)$ is continuous on $c < x < d$.

Then, $f(g(x))$ is continuous on $a < x < b$.

Let's look at how this might work in practice to show a complicated function is continuous.

Example 6.10   Show that

$$h(x) = \sqrt{x^2 + 1}$$

is continuous everywhere.

Solution:

Set $g(x) = x^2 + 1$ and $f(x) = \sqrt{x}$. Then $h(x) = f(g(x))$. We know $g(x)$ is continuous everywhere because it is a polynomial. We also have already shown that $g(x) > 0$ everywhere, and so the values it takes on are all positive. This means that $f(x)$ is continuous for every value that $g(x)$ can produce—which permits us to use the composition rule. We conclude that $h(x)$ is continuous everywhere.

◊

Example 6.11   Find the largest interval(s) on which

$$h(x) = \sqrt{x^2 - 1}$$

is continuous.

Solution:

The danger here is taking square roots of negative numbers. The composition rule may be applied, taking $g(x) = x^2 - 1$ and $f(x) = \sqrt{x}$ when

$$x^2 - 1 > 0.$$

This happens when x is at least one. So the maximal intervals of continuity of $h(x)$ are $(-\infty, -1)$ and $(1, \infty)$.

$\Diamond$

Example 6.12   Find the largest interval(s) on which

$$q(x) = \sqrt{x^2 - 4} \cdot \sqrt{\ln(x + 1)}$$

is continuous.

Solution:

The danger here also is taking square roots of negative numbers. Multiplication preserves continuity, but we need both halves of the product to be continuous for this to work. That means we need $x^2 - 4 \geq 0$ and $x + 1 \geq 1$. The solution to the first inequality is $(-\infty, -2) \cup (2, \infty)$, while the solution to the second is $(0, \infty)$.

Since *both* must be true to set up the product to yield a continuous function, we get that the negative range of the first solution is disallowed by the second and $(0, 2)$ is excluded from the second solution by the first, leaving us with the range $(2, \infty)$ where this function is continuous.

$\Diamond$

Example 6.13   Determine where $f(x) = \tan(x)$ is continuous.

Solution:

The tangent function is

$$\tan(x) = \frac{\sin(x)}{\cos(x)}$$

and we know that both sine and cosine are continuous everywhere. This means that the danger is division by zero. So the answer is $\cos(x) \neq 0$. The cosine is zero at odd multiples of $\pi/2$. So this means the tangent function is continuous on

$$\left\{ x : x \neq (2n + 1)\frac{\pi}{2} \right\}$$

where n is any whole number.

PROBLEMS

Problem 6.14   Use the formal definition of limits to prove the following limits.

1. $\lim\limits_{x \to 1} x + 3 = 4$

2. $\lim\limits_{x \to 2} 2x + 1 = 5$

3. $\lim\limits_{x \to -3} 5x + 3 = -12$

4. $\lim\limits_{x \to 5} x^2 = 25$

5. $\lim\limits_{x \to 1} x^2 = 1$

6. $\lim\limits_{x \to 2} x^3 = 8$

Problem 6.15   Suppose that $\lim\limits_{x \to 2} f(x) = 3$, $\lim\limits_{x \to 2} g(x) = -2$, and $\lim\limits_{x \to 2} h(x) = 7$.

Compute the following limits.

1. $\lim\limits_{x \to 2} f(x) + g(x)$

2. $\lim\limits_{x \to 2} 2.1 f(x) - 3.5 g(x)$

3. $\lim\limits_{x \to 2} f(x) + g(x) - h(x)$

4. $\lim\limits_{x \to 2} 3h(x) + 2f(x) + g(x)$

5. $\lim\limits_{x \to 2} \pi f(x) + e \cdot g(x)$

6. $\lim\limits_{x \to 2} h(x) - \pi g(x)$

Problem 6.16   Using the formal definition, show $\lim\limits_{x \to \infty} \dfrac{1}{x^3} = 0$.

Problem 6.17   Using the formal definition, show $\lim\limits_{x \to \infty} \dfrac{1}{e^x} = 0$.

Problem 6.18   Using the formal definition, show $\lim\limits_{x \to \infty} x^7 = \infty$.

Problem 6.19   Using the formal definition, show $\lim\limits_{x \to \infty} 5x + 6 = \infty$.

Problem 6.20   For each of the following functions, determine if the function is continuous at the value given. If the function is not continuous, say where the definition of continuity fails, e.g., "Not continuous because the function does not exist there."

1. $f(x) = 2x - 4$ at $x = 1$

2. $g(x) = \begin{cases} x^2 & x < 0 \\ x^3 & x \geq 0 \end{cases}$ at $x = 0$

3. $h(x) = \dfrac{2x}{x^2 - 16}$ at $x = 4$

4. $r(x) = \ln(x)$ at $x = -1$

5. $s(x) = \begin{cases} 5x + 11 & x < 3 \\ x^3 + x^2 - 10 & x \geq 3 \end{cases}$ at $x = 3$

6. $q(x) = \begin{cases} 5x + 11 & x < 2 \\ x^3 + x^2 - 10 & x \geq 2 \end{cases}$ at $x = 1$

Problem 6.21   Find all the largest possible intervals on which the following functions are continuous. Justify your results with a sentence or two.

1. $f(x) = \dfrac{x^2}{x^2 - 5}$

2. $g(x) = \sin^2(x) + 4\sin(x) + 5$

3. $h(x) = \dfrac{\sin(x)}{\cos^2(x) + 1}$

4. $r(x) = \sqrt{x^3 - x^2}$

5. $s(x) = \sqrt[3]{x^3 - x^2}$

6. $q(x) = \ln(\cos(x))$

Problem 6.22   If $p(x)$ is a polynomial, where is the function $f(x) = p(\cos(x))$ continuous?

6.2   THE SQUEEZE THEOREM AND THE MEAN VALUE THEOREM

Sometimes it is possible to compute a limit by bounding it by much easier limits. Consider trying to compute

$$\lim_{x \to \infty} \frac{\cos(x)}{x}$$

We already know that the infinite limits of sine and cosine do not exist, but this is different. Consider the graph of $\frac{\cos(x)}{x}$ shown in Figure 6.2.

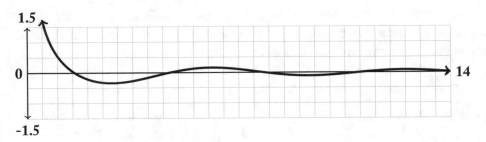

Figure 6.2: Graph of $\frac{\cos(x)}{x}$.

The cosine function is never farther from 0 than ± 1 but x grows without limit, making it look like the limit is zero. This lets us use a trick called the **squeeze theorem**.

Example 6.23   Compute:

$$\lim_{x \to \infty} \frac{\cos(x)}{x}$$

Solution:

Bound the limit and proceed.

$$-1 \leq \cos(x) \leq 1$$

$$\frac{-1}{x} \leq \frac{\cos(x)}{x} \leq \frac{1}{x} \qquad\qquad x > 0$$

$$\lim_{x \to \infty} \frac{-1}{x} \leq \lim_{x \to \infty} \frac{\cos(x)}{x} \leq \lim_{x \to \infty} \frac{1}{x}$$

$$0 \leq \lim_{x \to \infty} \frac{\cos(x)}{x} \leq 0$$

Since the limit we are interested in is *squeezed* between zero and zero, it follows that its value is also zero.

◊

The trick, in general, is to squeeze the limit of interest between two manageable limits. This trick is only useful when you can trap a limit in the fashion shown above. This most often happens when you have a ratio of functions that goes to zero. Let's formally state the result.

Knowledge Box 6.12

The Squeeze Theorem

Suppose that

$$f(x) \le g(x) \le h(x)$$

on an interval containing a. Then if

$$\lim_{x \to a} f(x) = L = \lim_{x \to a} h(x)$$

we may deduce that

$$\lim_{x \to a} g(x) = L.$$

This works if the interval is of the form (c, ∞) and the limit is as $x \to \infty$, as well.

Example 6.24 Compute $\lim_{x \to \infty} \dfrac{\sin(x) + \cos(x)}{3x + 2}$.

Solution:

We know that $-1 \le \sin(x), \cos(x) \le 1$.
This means that:

$$-2 \le \sin(x) + \cos(x) \le 2$$

$$\frac{-2}{3x + 2} \le \frac{\sin(x) + \cos(x)}{3x + 2} \le \frac{2}{3x + 2} \qquad x > 0$$

$$\lim_{x\to\infty} \frac{-2}{3x+2} \le \lim_{x\to\infty} \frac{\sin(x)+\cos(x)}{3x+2} \le \lim_{x\to\infty} \frac{2}{3x+2}$$

$$0 \le \lim_{x\to\infty} \frac{\sin(x)+\cos(x)}{3x+2} \le 0$$

Since the upper and lower bounding limits are 0 we may deduce

$$\lim_{x\to\infty} \frac{\sin(x)+\cos(x)}{3x+2} = 0$$

Knowledge Box 6.13

The Mean Value Theorem

Suppose that $f(x)$ is continuous and differentiable on the interval $a \le x \le b$. Then there is a value c with $a \le c \le b$ so that

$$\frac{f(b)-f(a)}{b-a} = f'(c).$$

Notice that

$$\frac{f(b)-f(a)}{b-a}$$

is the slope of the secant line to $f(x)$ on the interval $[a,b]$. What the mean value theorem says is that (at least) one point c in the interval $[a,b]$ has a tangent line to $f(x)$ that is parallel to the secant line—since $f'(c)$ is the slope of the tangent line at $x = c$. All that is happening is that we are applying Rolle's theorem to $f(x)$ after subtracting the secant line.

Example 6.25   Apply the mean value theorem to the function $f(x) = x^2$ on the interval $[-1, 2]$.

Solution:

The secant slope is

$$m = \frac{2^2 - (-1)^2}{2 - (-1)} = \frac{3}{3} = 1$$

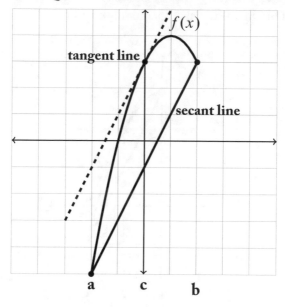

Figure 6.3: Illustration of the Mean Value Theorem.

Solve:

$$f'(x) = 1$$
$$2x = 1$$
$$x = \frac{1}{2}$$

So the value c predicted by the mean value theorem is $c = 1/2$.

◊

Example 6.26   Apply the mean value theorem to the function $f(x) = 3 - x^2$ on the interval $[0, 2]$ and draw the situation.

Solution:

The secant slope is $m = \dfrac{-1 - 3}{2 - 0} = \dfrac{-4}{2} = -2$. Solve:

$$f'(x) = -2$$
$$-2x = -2$$
$$x = 1$$

So the value c predicted by the mean value theorem is $c = 1$.

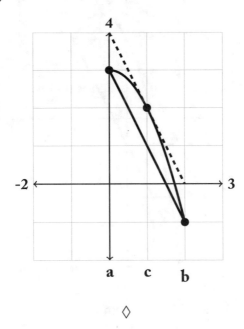

At this point the mean value theorem is a technical theorem. If you continue in your study of mathematics, it will turn out to be useful for proving facts about the nature of functions. For now, consider it a source of skill-building exercises.

PROBLEMS

Problem 6.27   Use the squeeze theorem to compute the following limits.

1. $\lim\limits_{x\to\infty} \dfrac{\cos(x)}{\ln(x)}$

2. $\lim\limits_{x\to\infty} \dfrac{\cos(x^2)}{x^2}$

3. $\lim\limits_{x\to\infty} \dfrac{\sin(x^2 + 1)}{x^3}$

4. $\lim\limits_{x\to\infty} \dfrac{\tan^{-1}(x)}{\sqrt{x}}$

5. $\lim\limits_{x\to\infty} \dfrac{\cos(2x)}{x + 1}$

6. $\lim\limits_{x\to\infty} \dfrac{\cos(\ln(x))}{x^2 + 1}$

7. $\lim\limits_{x\to\infty} \dfrac{\tan^{-1}(x)}{x^3 + x^2 + x + 1}$

8. $\lim\limits_{x\to 0} x \sin\left(\dfrac{1}{x}\right)$

9. $\lim\limits_{x\to 0} x \tan^{-1}\left(\dfrac{1}{x}\right)$

Problem 6.28   Find the value of c for the mean value theorem for the following functions and intervals $[a, b]$.

1. $f(x) = 4 - x^2$ on [-2,1]

2. $g(x) = x^3$ on [-1,1]

3. $h(x) = \ln(x)$ on [1,4]

4. $r(x) = \sqrt{x}$ on [0,9]

5. $s(x) = \sin(x)$ on $[0, \frac{\pi}{2}]$

6. $q(x) = \tan^{-1}(x)$ on [0,1]

Problem 6.29   For each part of Problem 6.28, carefully sketch the situation as in the examples in this section.

Problem 6.30   True or false:

$$\lim_{x \to 0} x \cdot \sin\left(\frac{1}{x}\right)$$

is zero. Explain your answer.

Problem 6.31   True or false:

$$\lim_{x \to 0} \sqrt[3]{x} \cdot \cos\left(\frac{1}{x}\right)$$

is zero. Explain your answer.

Problem 6.32   True or false:

$$\lim_{x \to 0} \sqrt{x} \cdot \sin\left(\frac{1}{x}\right)$$

exists. Explain your answer.

Problem 6.33   Construct an example of $f(x)$, a, and b so that the point c satisfying the mean value theorem is not unique.

Problem 6.34   Using the mean value theorem, prove that

$$|\sin(u) - \sin(v)| \le |u - v|$$

for any u, v.

Problem 6.35   Find two values a, b such that the secant line of $y = e^x$ has a slope of $m = 4$. Having done this, find the formula of the tangent line with the same slope.

Problem 6.36   Find two values a, b such that the secant line of $y = \ln(x)$ has a slope of $m = 2$. Having done this, find the formula of the tangent line with the same slope.

Problem 6.37   Find the maximum slope of any secant line of

$$y = \cos(x)$$

Problem 6.38   Suppose we take a car trip and have a digital logger record our velocity through-out the trip. Since the car is a mechanical device in the real world, we know that velocity as a function of time is a continuous function. If the trip took three hours, and we went 140 km, what is the slope at the mean value theorem point c for the logged function?

Problem 6.39   Show that there is a value $x = a$ so that the tangent line to $y = \cos(x)$ at $(a, \cos(a))$ goes through the point $(4, 3)$.

APPENDIX A

Useful Formulas

A.1 POWERS, LOGS, AND EXPONENTIALS

RULES FOR POWERS

- $a^{-n} = \dfrac{1}{a^n}$
- $\dfrac{a^n}{a^m} = a^{n-m}$
- $a^n \times a^m = a^{n+m}$
- $(a^n)^m = a^{n \times m}$

LOG AND EXPONENTIAL ALGEBRA

- $b^{\log_b(c)} = c$
- $\log_b(x^y) = y \cdot \log_b(x)$
- $\log_b(b^a) = a$
- $\log_b(xy) = \log_b(x) + \log_b(y)$
- $\log_c(x) = \dfrac{\log_b(x)}{\log_b(c)}$
- $\log_b\left(\dfrac{x}{y}\right) = \log_b(x) - \log_b(y)$
- If $\log_b(c) = a$, then $c = b^a$

A.2 TRIGONOMETRIC IDENTITIES

TRIG FUNCTION DEFINITIONS FROM SINE AND COSINE

- $\tan(\theta) = \dfrac{\sin(\theta)}{\cos(\theta)}$
- $\tan(\theta) = \dfrac{1}{\cot(\theta)}$
- $\csc(\theta) = \dfrac{1}{\sin(\theta)}$
- $\cot(\theta) = \dfrac{\cos(\theta)}{\sin(\theta)}$
- $\sec(\theta) = \dfrac{1}{\cos(\theta)}$

PERIODICITY IDENTITIES

- $\sin(x + 2\pi) = \sin(x)$
- $\tan(x) = -\cot\left(x - \frac{\pi}{2}\right)$
- $\sin(-x) = -\sin(x)$
- $\cos(x + 2\pi) = \cos(x)$
- $\sec(x) = \csc\left(x + \frac{\pi}{2}\right)$
- $\tan(x) = -\tan(x)$
- $\sin(x) = \cos\left(x - \frac{\pi}{2}\right)$
- $\cos(-x) = \cos(x)$
- $\sin(x + \pi) = -\sin(x)$

- $\cos(x + \pi) = -\cos(x)$ - $\tan(x + \pi) = \tan(x)$

THE PYTHAGOREAN IDENTITIES

- $\sin^2(\theta) + \cos^2(\theta) = 1$ - $\tan^2(\theta) + 1 = \sec^2(\theta)$ - $1 + \cot^2(\theta) = \csc^2(\theta)$

THE LAW OF SINES, THE LAW OF COSINES

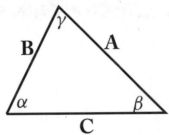

The Law of Sines

$$\frac{A}{\sin(\alpha)} = \frac{B}{\sin(\beta)} = \frac{C}{\sin(\gamma)}$$

The Law of Cosines

$$C^2 = A^2 + B^2 + 2AB \cdot \cos(\gamma)$$

The laws refer to the diagram.

SUM, DIFFERENCE, AND DOUBLE ANGLE

- $\sin(\alpha + \beta) = \sin(\alpha)\cos(\beta) + \sin(\beta)\cos(\alpha)$ - $\cos(2\theta) = \cos^2(\theta) - \sin^2(\theta)$

- $\cos(\alpha + \beta) = \cos(\alpha)\cos(\beta) - \sin(\alpha)\sin(\beta)$

- $\sin(\alpha - \beta) = \sin(\alpha)\cos(\beta) - \sin(\beta)\cos(\alpha)$ - $\cos^2(\theta/2) = \dfrac{1 + \cos(\theta)}{2}$

- $\cos(\alpha - \beta) = \sin(\alpha)\sin(\beta) + \cos(\alpha)\cos(\beta)$

- $\sin(2\theta) = 2\sin(\theta)\cos(\theta)$ - $\sin^2(\theta/2) = \dfrac{1 - \cos(\theta)}{2}$

A.3   SPEED OF FUNCTION GROWTH

- Logarithms grow faster than constants.

- Positive powers of x grow faster than logarithms.

- Larger positive powers of x grow faster than smaller positive. powers of x

- Exponentials (with positive exponents) grow faster than positive. powers of x.

- Exponentials with larger exponents grow faster than those with smaller exponents.

A.4 DERIVATIVE RULES

- If $f(x) = x^n$ then

$$f'(x) = nx^{n-1}$$

- $(f(x) + g(x))' = f'(x) + g'(x)$
- $(a \cdot f(x))' = a \cdot f'(x)$

- If $f(x) = \ln(x)$, then $f'(x) = \dfrac{1}{x}$

- If $f(x) = \log_b(x)$, then $f'(x) = \dfrac{1}{x \ln(b)}$

- If $f(x) = e^x$, then $f'(x) = e^x$
- If $f(x) = a^x$, then $f'(x) = \ln(a) \cdot a^x$
- $(\sin(x))' = \cos(x)$
- $(\cos(x))' = -\sin(x)$
- $(\tan(x))' = \sec^2(x)$
- $(\cot(x))' = -\csc^2(x)$

- $(\sec(x))' = \sec(x)\tan(x)$
- $(\csc(x))' = -\csc(x)\cot(x)$
- $\left(\sin^{-1}(x)\right)' = \dfrac{1}{\sqrt{1-x^2}}$
- $\left(\cos^{-1}(x)\right)' = \dfrac{-1}{\sqrt{1-x^2}}$
- $\left(\tan^{-1}(x)\right)' = \dfrac{1}{1+x^2}$
- $\left(\cot^{-1}(x)\right)' = \dfrac{-1}{1+x^2}$
- $\left(\sec^{-1}(x)\right)' = \dfrac{1}{|x|\sqrt{x^2-1}}$
- $\left(\csc^{-1}(x)\right)' = \dfrac{-1}{|x|\sqrt{x^2-1}}$

The product rule

$$(f(x) \cdot g(x))' = f(x)g'(x) + f'(x)g(x)$$

The quotient rule

$$\left(\frac{f(x)}{g(x)}\right)' = \frac{g(x)f'(x) - f(x)g'(x)}{g^2(x)}$$

The reciprocal rule

$$\left(\frac{1}{f(x)}\right)' = \frac{-f'(x)}{f^2(x)}$$

The chain rule

$$(f(g(x)))' = f'(g(x)) \cdot g'(x)$$

Author's Biography

DANIEL ASHLOCK

Daniel Ashlock is a Professor of Mathematics at the University of Guelph. He has a Ph.D. in Mathematics from the California Institute of Technology, 1990, and holds degrees in Computer Science and Mathematics from the University of Kansas, 1984. Dr. Ashlock has taught mathematics at levels from 7th grade through graduate school for four decades, starting at the age of 17. Over this time Dr. Ashlock has developed a number of ideas about how to help students overcome both fear and deficient preparation. This text, covering the mathematics portion of an integrated mathematics and physics course, has proven to be one of the more effective methods of helping students learn mathematics with physics serving as an ongoing anchor and example.

Index